ICE WORKS CONSTRUCTION GUIDES

Tunnelling

T. M. Megaw

Thomas Telford Limited, London
1982

CONTENTS

Other ICE Works Construction Guides available
Pile driving, W.A. Dawson
Earthworks, P.C. Horner
Trade unions in construction, Will Howie
Site handling equipment, J.R. Illingworth
Site investigation, A.D. Robb
Concrete: materials technology, J.F. Troy
Access scaffolding, C.J. Wilshere

Published by Thomas Telford Ltd, Telford House, PO Box 101, 26—34 Old Street, London EC1P 1JH

First published 1982

British Library Cataloguing in Publication Data
Megaw, T.M.
Tunnelling.—(ICE works construction guides)
I. Title II. Series
624.1'93 TA805
ISBN: 0 7277 0160 6

Typeset in Great Britain by MHL Typesetting Limited, Coventry
Printed and bound in Great Britain by H. Charlesworth & Co. Ltd, Huddersfield

1. GENERAL DESCRIPTION

The primary purpose in construction of a tunnel is to provide passage under or through a surface obstacle. Tunnelling techniques are also employed for such functions as excavation for underground power stations, provision of access to mines, storage and accommodation. Access and service tunnels in such cases do not differ greatly from transport tunnels. Excavation of large caverns, however, whether for turbines and generators or for oil or other storage does call for further development of rock excavation techniques, as does the construction of high pressure feed tunnels in hydroelectric schemes.

Location

The location and geology of a tunnel are major factors in determining the choice of construction method, the principal types being bored soft ground, bored rock, cut-and-cover, and submerged tube. Location may be broadly identified as mountain, hill, subaqueous or urban.

Tunnels through mountain ranges are almost invariably in rock, which may vary from hard and self-supporting to dangerously fragmented and unstable. Their usual functions are vehicular transport by rail or road, at acceptable gradients, and conveyance of water for irrigation, city supplies, dam bypasses or hydroelectric generation, including pumped storage schemes.

The term hill tunnel is here used for less severe terrain where a route for transport or water conveyance passes through a ridge or spur of high ground. This includes most of the canal tunnels. The ground may be hard rock or may be wholly or partly clay or other soft ground.

Subaqueous tunnels include those crossing beneath a river, estuary or other large body of water, mostly for railways, including metros, or roads but also for water supply and for pipes and cables. Also important are cooling water intakes and discharges for power generation or other industrial uses, and offshore sewage outfalls well beyond low tide. Various techniques of construction have their special applications.

Urban tunnels include more particularly metro systems, road and pedestrian underpasses, sewers, and pipe and cable subways, and also main line railways, city water supplies, and cooling water intakes and discharges. Much urban tunnelling is in soft ground because so many cities are sited on rivers and estuaries, but not infrequently rock is met at no great depth; cut-and-cover tunnelling is widely adopted because the shallowest practicable transport routes are preferred.

Function obviously governs dimensions and structure, although the primary structural requirement is that of ensuring permanent support of the surrounding ground. The earliest modern tunnels were for canals, in which the waterway must be truly level. Railway and highway tunnels are determined largely by limiting gradients and must conform to specified curves, having also adequate cross-section for vehicles to pass safely and freely. In tunnels conveying water the capacity is determined by cross-sectional area, smoothness and hydraulic gradient. For high pressure hydroelectric tunnels, watertightness is of particular importance, together with structural strength to carry both internal pressure when full and external pressure when empty. Sewer tunnels must usually be to a precise self-cleansing gradient and must be smooth and watertight. Pipe and cable tunnels can offer a useful alternative to burial in a trench, particularly for river crossings and in urban conditions.

Size and form

The range of size is from the smallest area in which a man can work, about 1½ m × 1 m, up to diameters such as 13.4 m OD for the Mersey Queensway highway tunnel or 32 m × 15 m for the Rove canal tunnel near Marseilles. Machinery halls in underground power stations may have spans upwards of 15 m. Lengths are from a few metres for a pipe heading, or a roadway through a rocky spur, up to about 50 km for the Channel Tunnel and Seikan Tunnel and 82.4 km for the Orange–Fish water tunnel. In cross-section the shape is normally a circle, arch, horseshoe or rectangle, the first being most usual for soft ground bored tunnels, some submerged tubes, and any tunnel bored by a full face machine, while an arched form is usually appropriate to rock excavation. This includes various shapes with an arched top, vertical or inclined sides, and a flat or arched invert. The rectangular form is almost invariably used for cut-and-cover, except in a transitional approach to a circular or arched type. It is also used in many submerged tubes.

Segmental linings, of cast iron or concrete, are common for circular tunnels, being fixed immediately following excavation in soft ground to provide permanent support as early as is practicable. In situ concrete is occasionally preferred where provision of full support is considered less urgent. Arched tunnels in rock are most usually lined with in situ concrete, often incorporating arch ribs which have been installed initially for immediate support. The permanent concrete lining may follow a considerable time and distance behind the excavation. In earlier days, for railway work, linings were almost always of brickwork—occasionally masonry if local stone was suitable. In some cases where rock is exceptionally sound no lining may be needed, except for possible local treatment of doubtful lengths with sprayed concrete. Rectangular cut-and-cover tunnels are almost always of reinforced concrete, except where shallow cover, or heavy loading, requires structural steel construction.

The structural lining may require further internal lining to provide a finish suitable for the tunnel's function. Examples are a smooth light-reflecting finish for highway tunnels or metro stations; minimum friction in tunnels conveying water; and resistance to erosion and to corrosive attack generally in sewer inverts.

2. TECHNOLOGY OF CONSTRUCTION

The three principal construction methods are boring, cut-and-cover and submerged tubes.

Boring is excavation from within, with the provision of necessary support and lining. It may be subdivided into soft ground tunnelling, where excavation is by hand or mechanical cutters and where immediate support is almost always essential; and rock tunnelling, where excavation is usually by drill and blast, but where powerful rock-cutting machines now offer an alternative.

In cut-and-cover, trenches are excavated from the surface and the structure is built in trench, with backfill over to restore the surface.

Submerged tubes are used for water crossings. Prefabricated tunnel units are built elsewhere and floated to site, where they are sunk and jointed in a trench excavated under water in the river or sea bed.

These methods are further discussed later, particularly bored tunnel techniques (sections 8–11). Cut-and-cover methods (section 12) are less specialised, having much in common with general deep foundation work. Submerged tubes (section 13) are in many ways a separate field.

3. SITE ORGANISATION

Quite elaborate organisation is essential in any large civil engineering project to ensure its efficient and economic execution. Much of what follows relates to any such project, but bored tunnelling calls for particularly detailed supervision and study of the ground conditions as exposed at the face, because construction procedures and details must be continuously adapted to changing ground.

For the purpose of elucidating the features and function essential in site organisation a large tunnel project is assumed, promoted by a public authority who employ a consulting engineer to design and supervise the work, and executed by a contractor. This is the most usual form in the UK, but other patterns of organisation are possible, particularly perhaps where the project is a small one, or tunnelling is incidental to other works. Execution of the work under the promoter's own engineers, or by direct labour, or under a 'package deal' are alternatives. Whatever the form, the site organisation must provide for safe and efficient execution, with adequate experienced supervision and with control of cost.

Figure 1 shows typical staffing at site for a large tunnel project, omitting clerical and administrative staff. For smaller tunnelling works numbers will be fewer, down to, perhaps, a single contractor's foreman and a part-time inspector, although shift working usually multiplies the numbers needed.

The importance at each level of having men of similar calibre and experience will readily be appreciated. In particular, mutual respect and understanding between Resident Engineer and Agent can contribute greatly to efficiency and smooth running and to the resolution of difficulties. It is essential that inspectors have practical experience and knowledge of tunnels such that their opinions and advice are of value to the engineering staff on both sides.

The separation of functions between Engineer and Contractor is less simple in tunnelling than in most other works, because design, construction and temporary works all interact with ground conditions, which are never entirely predictable. Close collaboration is therefore essential, particularly at site. There are bound to be disagreements from time to time at various levels, but the overriding consideration should always be to seek the best engineering

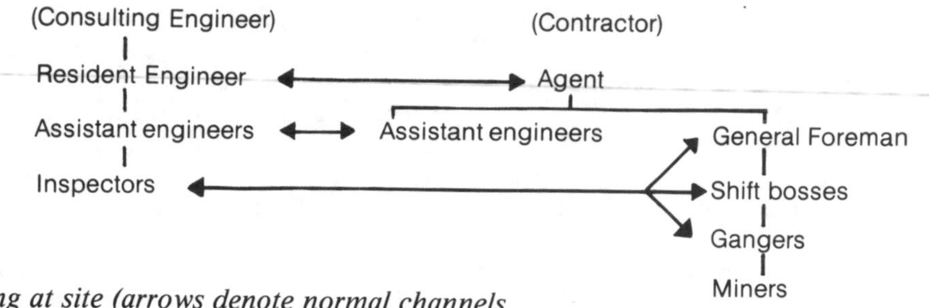

Fig. 1 Typical staffing at site (arrows denote normal channels of discussion between Engineer's and Contractor's staffs)

solution to each problem, even if the financial liabilities cannot be resolved in advance of execution. Ground forces do not remain static while discussions go on, and safety may be imperilled by delay.

4. SAFETY

Safety in tunnelling is the subject of British Standard BS 6164.[13] There are special hazards in tunnelling because of the nature of the work, the confined space and problems of access. There should be contingency planning in advance of construction to ensure that even unexpected hazards do not become catastrophes.

Natural hazards arising from the ground conditions include collapse, inundation and gas. Hazards inherent in the operating procedure include restricted access, limited clearances from equipment, handling and use of explosives, limitations of lighting, noise, dust, fumes, electrical hazards in cramped and wet conditions, fire, and compressed air working.

It is not possible here to do much more than draw attention to the dangers, but the most important safety measure is to be aware of them at all times and to know and take the appropriate action before serious trouble develops.

Collapse of the ground is a risk inherent in tunnelling, because the act of excavation from inside and beneath removes support. In soft ground the danger of progressive loosening and flow is obvious, and may be greatly exacerbated by the presence of water. In rock, the shock of blasting and removal of support may also initiate a substantial collapse, possibly attributable to unexpected joint patterns, faults, a change in stratum or the presence of water. The first line of defence is the limiting of excavated area and the availability of materials for immediate support.

Inundation at the face is an ever-present hazard in subaqueous tunnelling, or in any conditions where a very large volume of groundwater may be encountered. In shield tunnelling a protective diaphragm may be provided, or bulkheads with access doors may be built. Other protective measures are the use of compressed air, and ground treatment by injection of grouts or by freezing. Protection against flooding from access shafts or other openings, such as pipe or cable headings, should also be considered carefully. The most usual hazard of gas is of methane, known in mining as 'fire damp', which is readily explosive when mixed with air. Its incidence is most to be feared in the vicinity of coal seams but it may also be generated in other organic strata such as peat. Frequent testing, and ample ventilation, together with precautions against fire, hot spots and sparks, whether electrical or mechanical, are the appropriate actions. Other gas hazards are from carbon dioxide and carbon monoxide, and, more rarely, sulphur dioxide. Petroleum gases may also be encountered in some areas. 'Deoxygenated air' may arise in tunnels where air enters the tunnel from an organic stratum in which oxidation has taken place. The air may be drawn in by a drop in barometric pressure or may be driven in where compressed air working is in progress in the vicinity. In sewer tunnels which have been in use a special hazard is hydrogen sulphide.

Restricted access, as by a single shaft and access tunnel, may limit severely rescue operations. Limited clearances from moving plant such as locomotives or excavators demand great care from individuals, for whom strict safety rules should be made and applied. Care in handling and use of explosives is an obvious necessity. British Standard 5607[14] has a section relating to tunnel work. Dust, from the breaking of rock or from drying out of clay, should be suppressed by water sprays as near as possible to its point of generation. Silica dust is a particular hazard to lungs. Lighting within a tunnel should be of a high standard, with special attention being given to working areas. As with other services there must be continual extension as the tunnel advances. Noise is now being recognised as a hazard, not merely in terms of possible permanent damage to hearing, but also because it makes communication difficult and can distract attention from other dangers. Electrical supplies and plant need careful attention in the typically damp conditions. Earthing may need particular study, and cables passing along a tunnel require protection against mechanical damage and against submersion in water. In the presence of methane, flameproof electrical equipment may be demanded.

Fire in a tunnel may produce carbon monoxide, smoke may make access or escape very difficult, and combustion of timber supports may initiate collapse. Use of burning and welding gear must be carefully controlled. Compressed air working gives rise to physiological risks, covered by special regulations. Physical risks also occur: a 'blow' at the face will result in loss of air pressure, and there is a special fire hazard due to the greater availability of oxygen.

Safety on a job is best ensured by individuals at all levels having a responsible attitude, and by there being a system to ensure that hazards are reported and that appropriate action is taken. A large site organisation will probably include a named safety officer and a safety committee.

5. SITE SURVEYS

In the early planning stages, traffic studies and investigations into economic costs and benefits will be made, and necessary powers for land acquisition and use will be obtained, possibly by a Private Act in Parliament. Planning will be based on existing records and topographical and geological information, such as Ordnance Survey and Geological Survey maps, but as the project is brought to the contract stage much more precise detail becomes essential.

In open country the survey will be based on an accurate triangulation network, by which reference points and base lines are established, usually by both Engineer and Contractor, for setting out and control of the construction. Modern electronic distance meters (EDMs) replace the calibrated steel band for long lengths, and make possible length checks on any line in a triangulated network, with special advantages over rough ground. It is even possible to replace triangulation, wholly or partly, by trilateration. Permanent monuments, not subject to later disturbance, should be constructed to define the tunnel centre-line.

In urban conditions, the network is more likely to be a polygonal traverse in which all angles and distances are measured, and in which details are incorporated by offset measurements from main or subsidiary survey lines. Because the steel band allows intermediate distances to be noted, and because lengths of sides of the polygon tend to be short, it is not replaced by the EDM for ground-level work. The EDM may, however, be used for a network at roof top level.

Aerial survey can very usefully furnish detail both in open country and in cities, but must be controlled by a ground-level network.

Levels over the site are an important feature of the survey, and a network should be established with precision, for use during construction and also, particularly in towns, for measurement and control of any settlement.

The essence of all bored tunnel survey is great precision, because in driving the tunnel, lines must be extended continuously as the face advances, usually without possibility of any direct check on position.

The geology, geotechnology and hydrology of the site must also be the subject of preliminary and continuing study. Data from geological maps and surface inspection have to be confirmed and supplemented by the making of boreholes to determine the characteristics of the ground at depth and its detailed geological structure (Fig. 2). The more boreholes sunk for this purpose the better. It is usually most efficient to start with a limited number of bores in order to choose the most favourable alignment and depth for the tunnel, and to follow with further boreholes closer to the selected line to fill in detail and clarify uncertainties, even during the progress of the contract. It is important to ascertain the nature of the ground beneath the tunnel and, for this purpose, sufficient of the boreholes should extend to a level well below the tunnel invert. Undisturbed samples and rock cores from borings are required for tests and studies of the ground, in terms of soil mechanics or rock mechanics as appropriate. The relevant properties to be tested, and the number and extent of the boreholes for the particular tunnel must be selected

Fig. 2. Drilling rig and tools (Engineering Laboratory Equipment Ltd): (left) mobile drilling rig, with percussion and auger capability; (right) tools for use with drilling rig—(1) drilling bit; (2) auger bit for soft ground; (3) and (4) chisels for breaking out rock; (5) clay corer; (6) shell for cleaning out hole after drilling; (7)–(9) gravelling, auger and serrated shoes for use with shell

beforehand, but should be subject to modification if available resources are to be used to the best advantage. The study should also take account of groundwater, finding water table levels, permeability, acidity and other features; surface drainage, flooding, tides and even rainfall may also be directly relevant.

Other subjects for preliminary investigation are such matters as road and rail access, and resources of labour and materials.

6. SITE PREPARATIONS

The first work at site, apart from survey, is likely to be the construction of access roads, assembly of plant, and provision of accommodation, all located with proper regard to the permanent structures. Sinking of one or more access shafts (described in section 7) may be the beginning of construction; such shafts are often for permanent access or ventilation, but sometimes temporary only.

Arrangements for disposal of the excavated spoil are important to efficient working. In open country, on-site disposal may be practicable, or the material may be utilised for building of embankments. In cities, hopper storage at the shaft head is essential if transport to a tip is restricted by street traffic congestion during much of the day.

The assembly and siting of plant and the arrangements for electricity supply are important to the programme for construction. Drainage must be planned, including pumped water from the tunnel workings, possibly with settling tanks to prevent silting of drains. For special items such as tunnel shields or tunnel boring machines, there may be a long lead time between order and delivery, and where segmental linings are to be used the adequacy and timing of supplies must be ensured. Cranes, hoists, conveyors, haulage equipment, air compressors, pumps etc. have to be obtained and brought to site. Suitable foundations and sheds must be constructed in positions clear of the permanent work. For a large project electrical supplies are likely to be at high voltage, and to require heavy switchgear, transformers and cabling. For compressed air tunnelling, in addition to provision of large low pressure air compressors, absolute security of air supply may be vital, demanding duplication of electrical supply and also standby diesel compressors.

7. SHAFTS

The most useful shaft diameter is 6 m or more, giving room for a hoist and access ladders and for crane loads, but 4 m is quite practicable for many purposes. Although the tunnel may be sited at a depth where firm ground is expected, most shafts will be sunk through softer and looser surface strata, very frequently waterbearing. Various shaft linings and sinking methods may be adopted, depending in part on the ultimate use of the shaft. A timbered pit, or sheet piling with timber or steel framing, can be used for sinking down to firm ground, below which a permanent circular shaft may be constructed, lined with

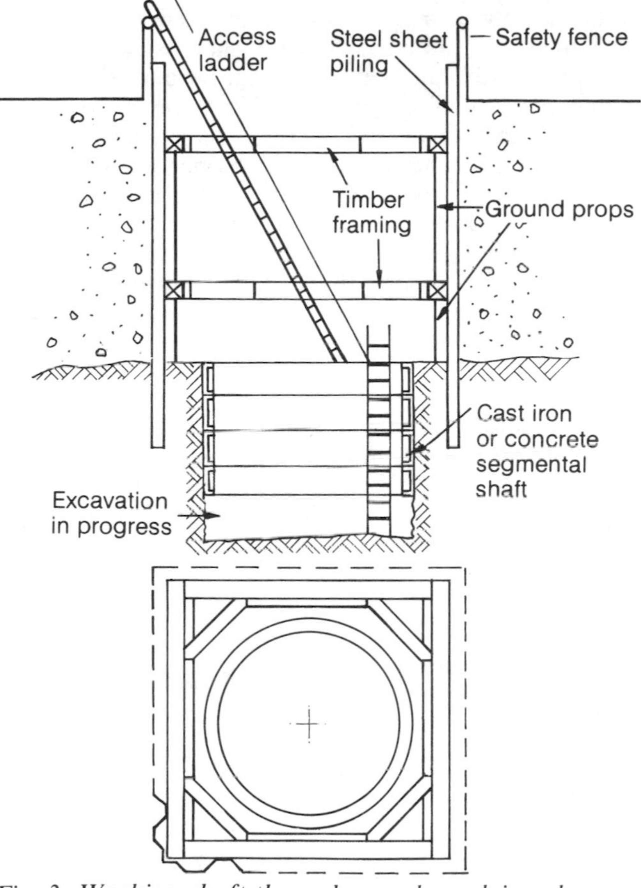

Fig. 3. Working shaft through gravel overlying clay

Fig. 4. Ventilation shaft with concrete walls cast in slurry-filled trenches

cast iron or concrete segments (Fig. 3), compressed air working being employed if necessary. Concrete walls built by slurry trench methods (Fig. 4) or close-fitting bored concrete piles are other possibilities. Caissons sunk in compressed air may be advantageous in difficult ground, particularly where they can be designed to embody complex permanent structures (Fig. 5).

The bottom of any access shaft requires particular care in two respects. First, it must provide for the breakout for an access tunnel or the main tunnel, for which a 'tunnel eye' may be embodied in the shaft structure, with appropriate stiffening and strutting. Second, the bottom of the access shaft requires the deepest excavation. The shaft bottom is usually sealed with a concrete slab, embodying a sump and other facilities for drainage and pumping. In excavating to the deepest level there is frequently a danger of ground heave from plastic flow in a bed of clay, or uplift from the pressure of water in a permeable stratum below the base. Preventive action may be to excavate the last layer in narrow strips, refilled immediately with concrete, or to relieve the excess water pressure by sinking a small borehole through the bottom. Safety at the top of a shaft usually requires a solid curb and a railing, and appropriate precautions against flooding. A crane usually serves such a shaft: heavy loading of the ground whether from the crane or stacked material in the immediate vicinity of the shaft must be avoided, or proper structural provision must be ensured.

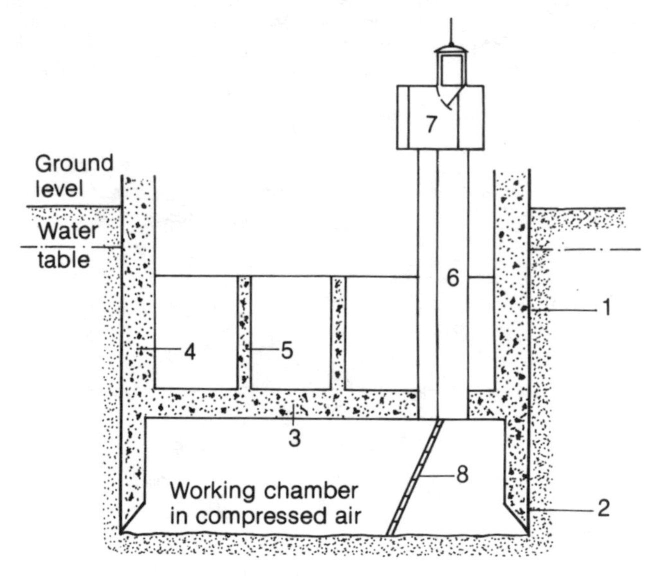

1. Steel shell
2. Steel cutting edge
3. Air tight deck
4. Walls built progressively during sinking
5. Internal structure as required
6. Steel 'figure 8' access shaft with—
7. Air locks for men and materials
8. Access ladder

Fig. 5. Compressed air caisson

8. SOFT GROUND TUNNELLING

For all bored tunnels the essential factors are

(a) excavation from within and below, with removal and disposal of spoil;
(b) immediate support as necessary to roof and face;
(c) permanent support and lining;
(d) management of water;
(e) survey and control.

Types of ground

Whether in soft ground or rock, the principles of advancing in safe steps and providing appropriate ground support govern operations, and although separately discussed, the techniques have many factors in common.

Soft ground may be described as that which can be dug out and which is not self-supporting for more than a brief period. It corresponds to the geologists' unconsolidated sediments. Most soft ground tunnelling is at relatively shallow depth, in urban situations. The typical sediments may be classified, in order of grain size, as clay, silt, and gravel, to which may possibly be added soft chalk. Further subdivision to any degree is possible in terms of mineral and organic content, grain size and grading and the whole range of soil mechanics properties. A most important feature is the degree of consolidation and permeability of the ground and the presence of groundwater.

In all soft ground work the techniques of excavation, support and water management must be closely integrated, the system adopted being adaptable to modifications as the ground inevitably changes.

Clay is a very fine grained sediment and may range from near liquid to very stiff according to its natural water content and degree of consolidation. Except where fissured, it is of very low permeability, so that it forms a barrier to flowing water. Very soft clay demands immediate and close-timbered support concurrent with excavation. Stiff clay may stand for a considerable time with minimum support, but the risk of slow and almost imperceptible plastic deformation, and consequent surface settlement, must be guarded against.

Silt is one of the most dangerous types of ground for tunnelling. It is of coarser grain size than clay and is more permeable to water. When damp it can have cohesion which makes it appear self-supporting, but this only applies over a very limited range of moisture content, and its hazards arise from the nearly complete loss of cohesion when it is either drier or wetter than its optimum, together with its fine grain size which allows it to escape through very small gaps in support timber.

Organic silts are of low permeability but high compressibility.

Sands and gravels are classified by grain size and grading (Fig. 6). They lack cohesion, except to a very limited degree when slightly damp, and their permeability increases with grain size. Clean sands and gravels run readily when dry and when saturated, and in tunnelling demand skill and care in the combined operation of excavation and timbering, most particularly in sub-aqueous conditions. If there is appreciable clay content they may develop very useful cohesive strength.

Chalk varies from soft uniform rock with a substantial

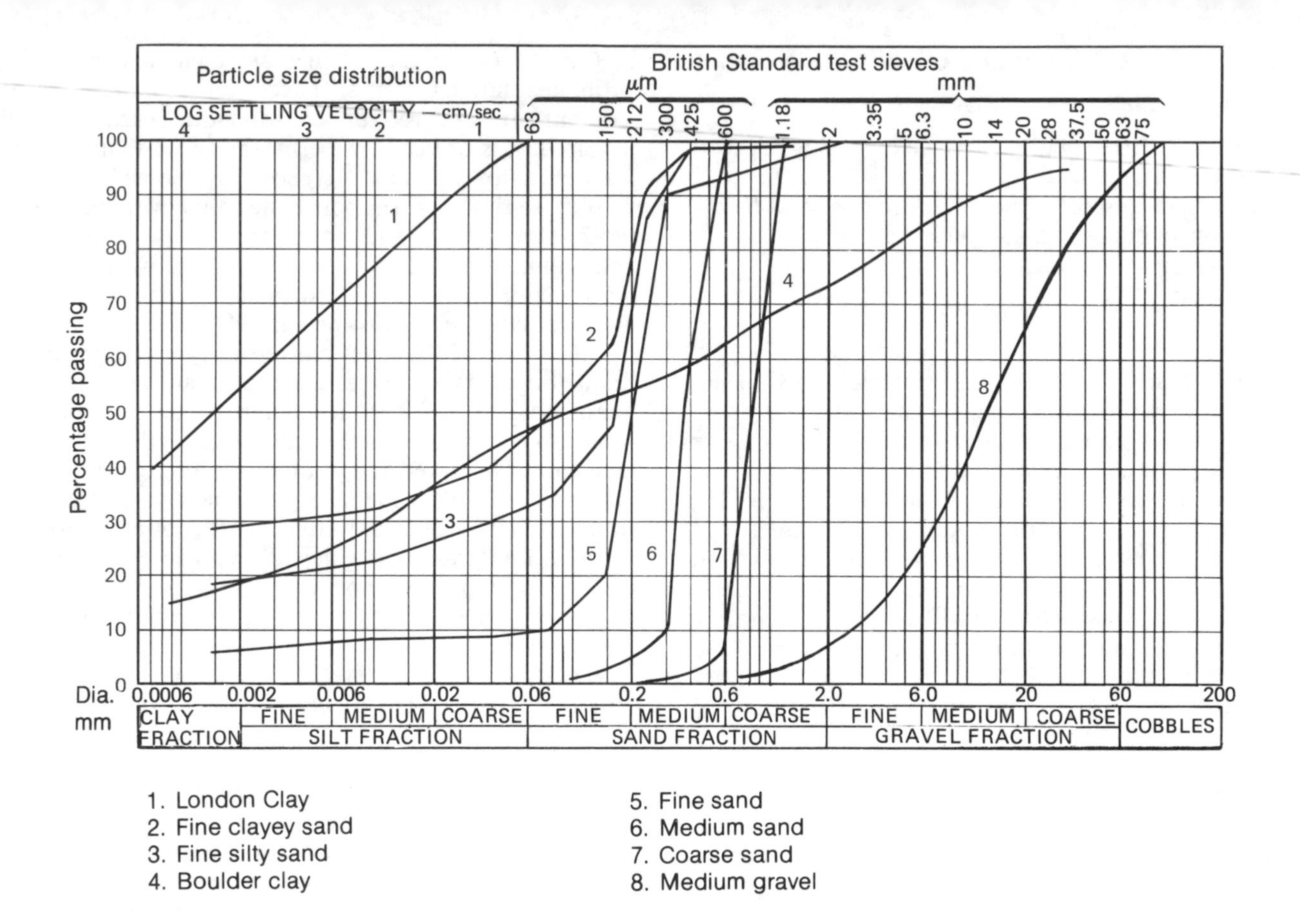

Fig. 6. Soil grading curves

clay content to a hard limestone, fissured and with bands of flint: excavation and support techniques must be adapted accordingly.

Timbered headings

The term 'heading' was originally applied to a small tunnel driven ahead of the main excavation, timbering and (brick) lining. In UK practice the word is now used for any small timbered tunnel, wherever sited. It is so used below. In US practice it is sometimes differently applied, referring primarily to the area where the main face is being advanced.

A typical timbered heading may be of the order of 2 m high × 1 m wide, or even, for short lengths, as small as 1 m × 0.6 m. The pattern of timbering is shown in Fig. 7. Similar procedures apply when timber is replaced by steel joist or channel frames with timber or steel boards.

The description below is for ground needing particular care, either because of its weakness or because of possible consequences of subsidence. In different conditions details may be modified.

The sequence of operations in advancing such a heading in 'settings' of about 1.2 m, from an already completed length, with the last completed frame in position at the face as shown in Fig. 7, is as follows:

(a) Remove any face boards as required to excavate top, and fix new head boards spanning from old headtree to temporary prop at new face.
(b) Complete excavation, fix and wedge up new headtree and sidetrees and side boards, also any necessary face boards and also footblocks or sill under sidetrees. Sidetrees are set with 'sprag' (i.e., sloping at about 1 in 6 outwards from vertical).
(c) Ensure that all is tightly wedged and packed, and grouted if necessary. Head boards may be 'buttered' with cement mortar to ensure better support of the ground.
(d) Repeat above sequence.

This exhibits the whole essence of soft ground tunnelling up to the immediate support stage. It will be clear that each step can be adapted to the ground as it is encountered, the time available between excavation and support being critical. For very soft or loose ground further precautions may be necessary such as close-timbering the top, face and sides, and in extreme conditions driving the boards as horizontal piles in advance of excavation. A fully piled heading makes it possible to advance through almost any soft ground. Progressively greater skill and experience in timbering are necessary for such ground.

The presence of water obviously adds to the difficulty and to the care and skill required. In fine sand or silt there is a danger of fine material being washed in through any gaps: accurate closely fitting timbering is necessary, assisted by clay 'pugging' of cracks.

In a larger tunnel the problems of support of top, face and sides become rapidly more severe, but the adoption of segmental linings forming circular rings erected immediately behind the excavated face allows advance to be made in short steps, with temporary timbering, in the face and top, reused as each ring is erected.

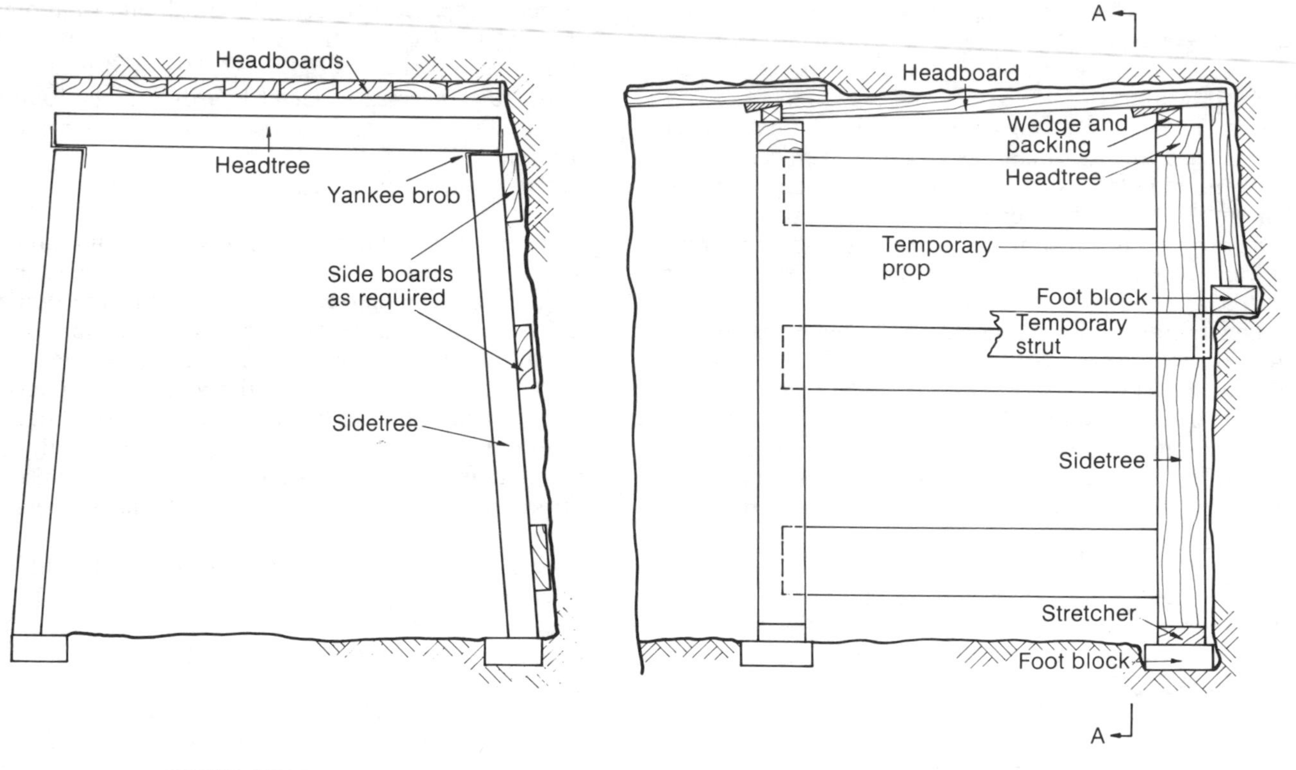

Fig. 7. Timbered heading

Shield tunnels

Larger tunnels cannot simply be small ones scaled up. There is a disproportionate increase of difficulty, particularly in problems of support. Even if the step ahead is limited to the same length as in the small heading, the increased face area must be supported across wider spans using heavier timbers. The increased volume to be excavated and the heavier timbering require a longer time for the cycle of operations, during which stresses and displacements in the ground develop.

This leads to the concept of 'stand-up time'. The term was introduced into soft ground tunnelling by Terzaghi who, in 1950, described it as the time elapsing between exposure of the roof and noticeable movement of the ground above. It is therefore a property of the ground, and also a function of the span across the tunnel and the span between the last fixed supporting frame or arch and the face. The level of the water table is important. Charts and tables have been produced indicating times ranging from less than half a minute in running ground below the water table, where the span must obviously be kept down to very small dimensions, up to two hours or more in fine sand with clay providing some cohesion, and a day or

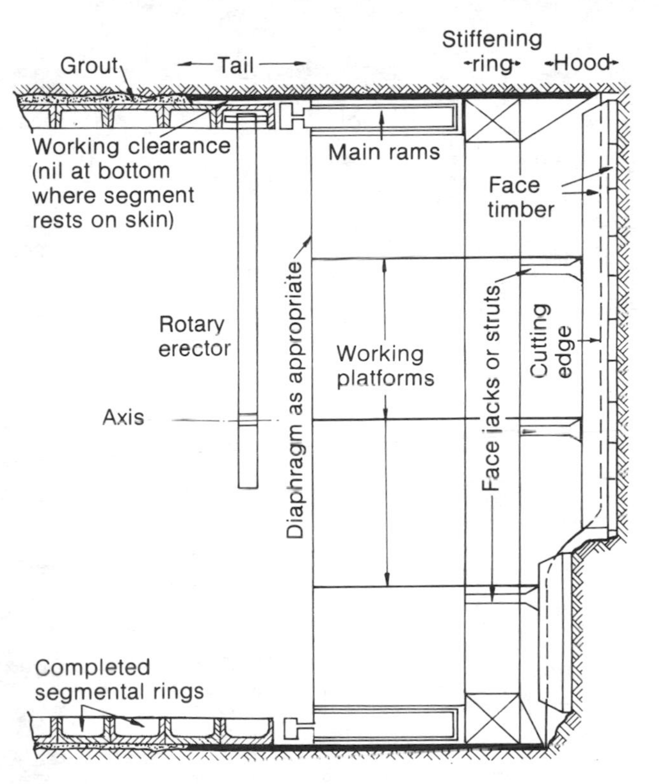

Fig. 8 (right). Large Greathead shield in use: the erection of a ring within the tail has just been completed; the next operation is the advance of the face by excavating from the top downwards, moving the timbers forward; the shield is then shoved forward by the length of one ring by extending the main rams against the completed lining—controlled release of the face struts is necessary during this movement; grouting of the last completed ring follows, together with erection of a new ring

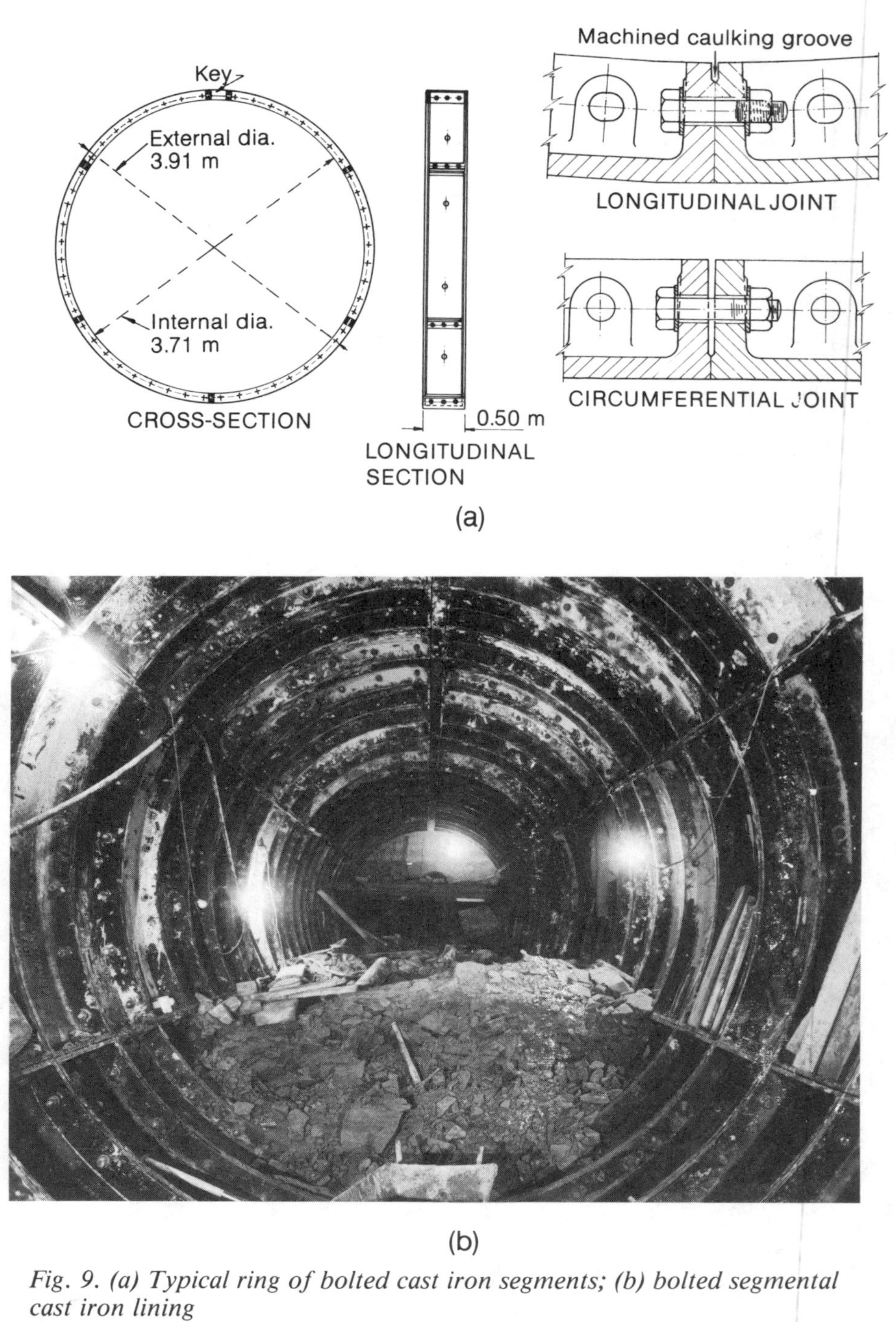

Fig. 9. (a) Typical ring of bolted cast iron segments; (b) bolted segmental cast iron lining

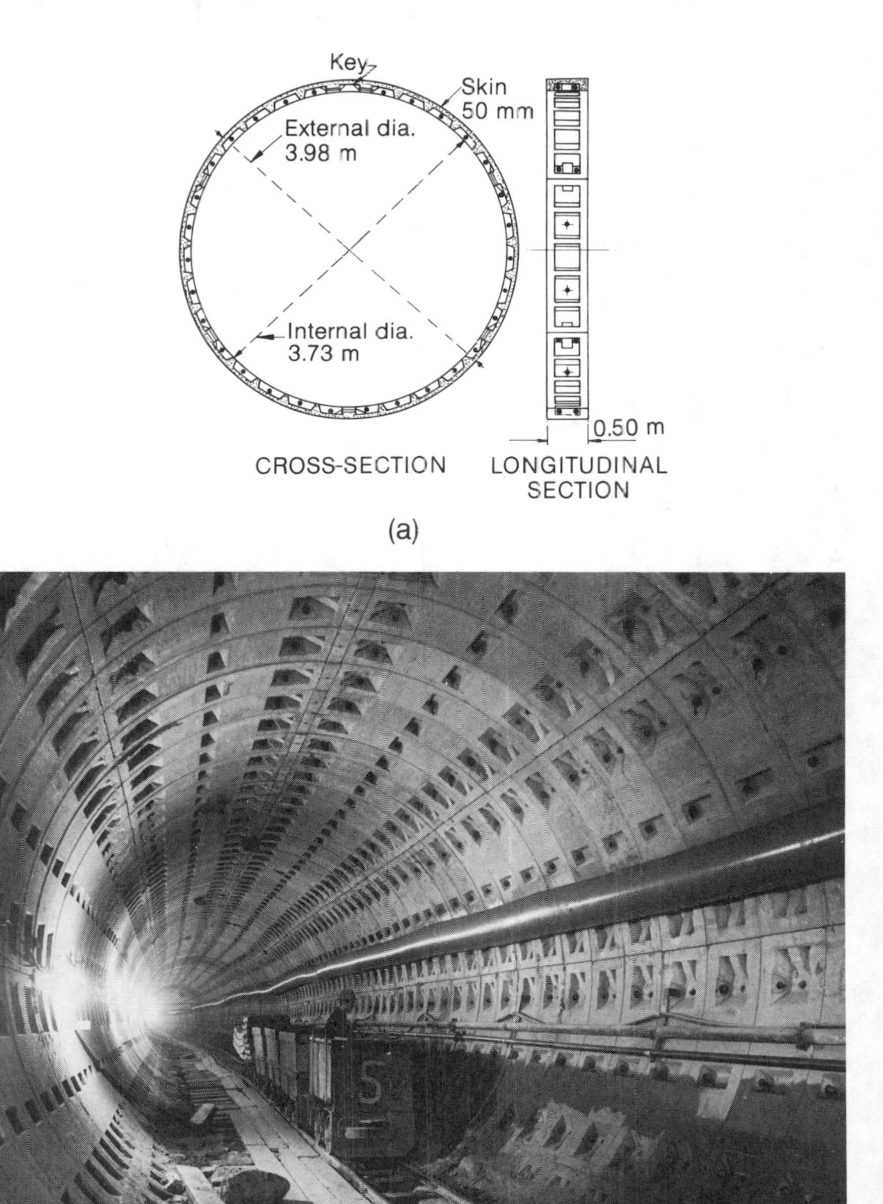

Fig. 10. Precast concrete segmental linings: (a) original London Transport pattern; (b) as used in Toronto subway

more in firm well bound sand or gravel with clay. The span, correspondingly, ranges up to 2–3 m.

The Greathead shield, first used in conjunction with cast iron segmental lining for the small Tower Subway in 1869, is invaluable in meeting these difficulties. In its simplest form it comprises a cylindrical shell with a stiffening ring carrying face jacks used to propel it forward. The ground ahead of the face is excavated for a length of one ring and the shield is shoved forward to trim and support the periphery of the excavation. In larger tunnels a timbered top heading is sometimes excavated a few rings ahead of the main face. The shield skin has a cutting edge at its leading end and, normally, a tail skin behind, within the protection of which a new ring of lining is built. The extent to which face support timbering is necessary depends on the ground and the timing. Fig. 8 shows the sequence for a large shield. In very soft ground or water-bearing gravel it may be necessary to close-timber the face and to remove and advance the support one board at a time. In such circumstances a diaphragm with a sliding or other door will be fitted across the shield, and face jacks bearing on cross-beams will be provided to support the main face timbers. In good stiff clay little or no face timber may be necessary in a small tunnel when work is proceeding steadily and rapidly, but face timber will invariably be required if there is a delay or shutdown for any reason.

When a ring of segments is built within the tail skin, a void between it and the excavated ground is left after the next 'shove'. This is immediately filled with cement grout injected under pressure.

In stiff clays or similar ground having a sufficient stand-up time the tail skin may be dispensed with and flexible jointed segmental linings may be adopted.

Segmental linings

Except for timbered headings and jacked pipes, segmental linings erected immediately behind a shield are almost always used in soft ground. The standard material for many years was cast iron (Fig. 9); reinforced concrete of similar pattern (Fig. 10) was introduced in the 1930s and it has now been developed in many different patterns. Some flexibility to conform to strain deformation of the ground is an important feature which reduces bending stress and

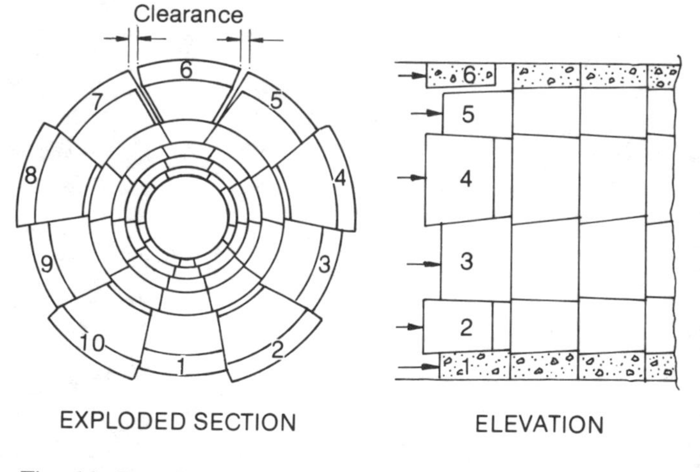

Fig. 11. Don Seg lining

cracking. There are two principal types: flanged and bolted segmental rings, which may be built without a shield, or within the tail skin, and which are embedded solidly within the ground by pressure-grouting; and flexibly jointed rings, which are wedged out directly against the excavated perimeter. Some are wedged out by use of longitudinally tapered segments (Fig. 11) and some by direct jacking when the ring, with parallel joints, is assembled.

Except for severe conditions of stress, or sometimes in waterbearing ground, concrete has now largely superseded the more costly cast iron.

Tunnel boring machines

Tunnel boring machines (TBMs) for soft ground almost invariably operate in conjunction with a Greathead shield and segmental lining. There are two principal types: the full face rotary excavator and the independent excavator. The rotary excavator revolving within the circular shield may be carried on a central shaft or on peripheral roller bearings, and is most usually powered by hydraulic or electric motors operating through an annular gear. Cutting tools are fitted to radial spokes and the spoil is picked up by scoops which discharge it to a conveyor. A variant type has an oscillating head. The independent excavator may be an independently mounted boom cutter, based on the mining road header, or may be a bucket excavator. There are also claw-armed digger shields, with semi-circular cutter arms mounted on an axis close to a diameter of the face, scooping out a hemisphere of ground ahead. A further development for waterbearing ground is typified by the bentonite shield, where the face is closed off by a bulkhead, the space ahead being charged with bentonite slurry. A cutter head rotates within this chamber and discharges spoil and slurry through a control gate. The slurry in the chamber is continuously replenished and its pressure provides support for the face. Bentonite discharged with the spoil is recovered for recirculation. Other types of slurry shield and earth pressure balance shield have been developed.

The economic use of a TBM depends on having a drive sufficiently long and uniform to offset the high initial cost of the equipment, and having the time and length of tunnel necessary to install and bring into full production all the usual equipment for handling and erecting segments, and for grouting, together with spoil-handling conveyors and loaders.

For very small tunnels the Mini-tunnel system has been devised. Another technique is that of pipe jacking, in which the leading length of pipe acts as a shield and the whole length of pipeline is jacked forward from an initial pit as excavation proceeds at the face, new pipe lengths being added in the pit. A very recent development is the Unitunnel, in which the jacking force is minimised by progressive use of pneumatic joint rings between pipes, which allow a creeping movement forwards two rings at a time

Management of water

The management of water in any soft ground tunnel is likely to be an important factor in safety and productivity. Means for ground treatment to reduce inflow are discussed below, but they are unlikely to exclude water completely, and provision for handling it must be made accordingly. There is advantage in driving uphill in a wet tunnel because water entering at the face can flow back naturally, but a downhill drive may be essential, as in a subaqueous crossing. In this, water accumulates in the invert at the face and adds to the difficulty of cleaning out the invert, erecting lining segments, and grouting. A small sump very close to the face is required, with reliable pumping equipment which can be moved forward progressively; water is pumped back to a larger and more permanent sump, whence in turn it is pumped back to and up the working shaft. Any interruptions, by breakdown or choking of pumps or pipes, may cause troublesome flooding at the face and disruption of output. Pumping equipment must be chosen to handle the inevitable muddy water and spills of cement grout, and adequate screens and filters must be provided to protect it against debris.

In wet ground (as already remarked), timbering of roof and face may have to be so close as to prevent the washing in of fine material from the ground.

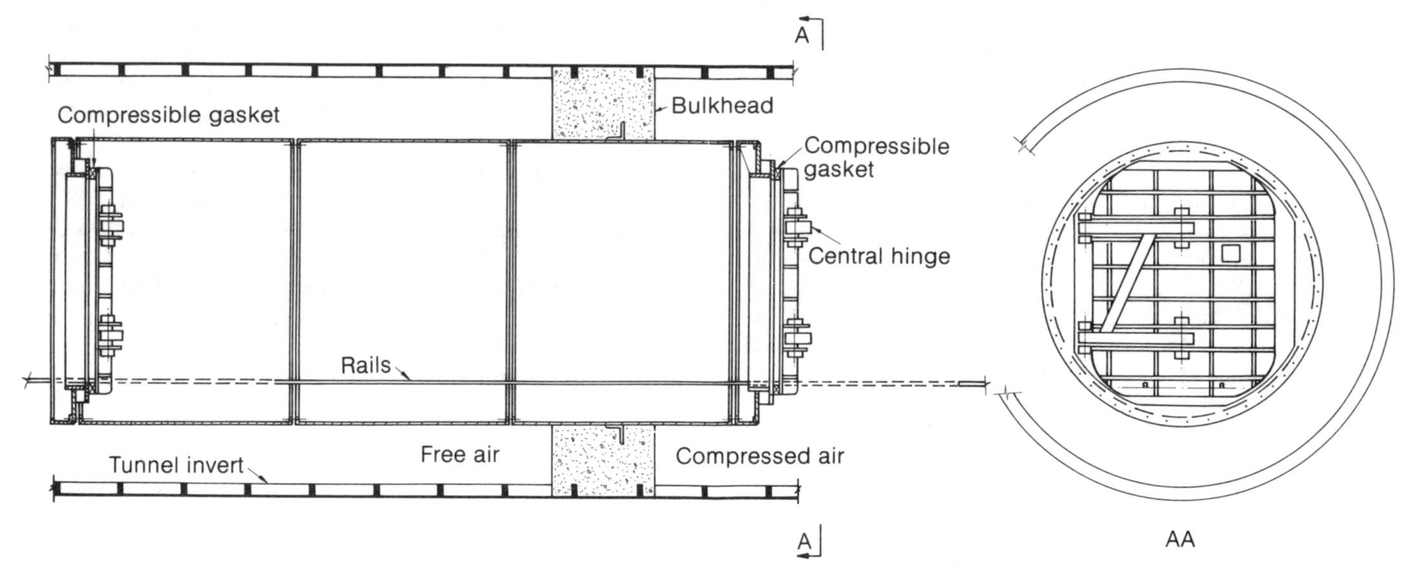

Fig. 12. Boiler lock in tunnel

9. ANCILLARY OPERATIONS IN SOFT GROUND

Compressed air

Compressed air in tunnelling serves to hold back groundwater at the face and to provide useful support. The air must of course be supplied from compressors and retained by airtight bulkheads, or decks to shafts, through which men and materials pass by means of air locks (Fig. 12). This use of compressed air was developed towards the end of the 19th century in the Hudson River Tunnel, the City and South London Railway, and the Blackwall Tunnel. There are two aspects to be considered: the physical actions and reactions in the tunnel and surrounding ground, and the physiological effects on men working in it.

The use of compressed air makes possible tunnelling in ground where it would otherwise be impracticable, and can greatly increase safety and efficiency in less extreme conditions. It is used in waterbearing silts, sands, gravels and fissured material to counteract the pore water at the face and in the top, and also as a precautionary measure to provide support in clay where the impermeable cover provided by the clay is of inadequate thickness, or may be breached altogether.

In open ground, measures may be needed to reduce air loss through the face and completed lengths of tunnel.

The appropriate working pressure is determined in relation to the hydrostatic water pressure at the working face. Equilibrium is not attainable because air pressure is uniform, whereas water pressure increases from top to bottom (Fig. 13). Judgement based on experience is necessary in selecting a balancing level in the face. Above this level air pressure exceeds hydrostatic pressure and air escapes: at lower levels water pressure is greater and water enters. Excess air pressure results in high air losses, disturbance of ground and danger of a 'blow', and also lengthens decompression time for workers. Deficiency in pressure increases the inflow of water, possibly carrying in

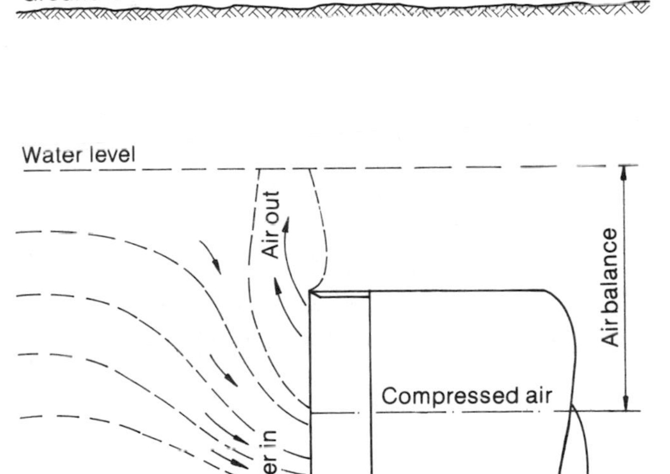

Fig. 13. Air balance in tunnel

fine material, and increases the loading to be carried by timbering at the face and top.

Working air pressure is described, not as absolute pressure, but as the excess over atmospheric pressure, or 'gauge pressure'. Pressures up to about one atmosphere (about 1 bar in SI units, or 14 lbf/in^2 gauge) meet most requirements for shallower tunnels down to 10 m below water table.

The plant and equipment essential to compressed air working, on any but a small low pressure job, will include the items shown in Table 1. High costs are inevitable in the provision and maintenance of equipment, in direct operating costs, and in problems of access for labour and materials, non-productive time in decompression, and appropriate rates of pay, together with skilled supervision. Nevertheless, security against collapse and against subsidence of adjacent structures often justifies precautionary application of compressed air.

Table 1. Plant and equipment for compressed air working

Tunnel bulkhead or air deck
Air locks:
Man lock
Materials lock
Medical lock (above 1 bar pressure)
Low pressure compressor installation:
Compressors
Standby compressors, up to 100%
Intake filters
After coolers
Oil scrubbers
Air receivers
Duplicate mains to tunnel
Control valves, gauges, telephones etc.

The physiological problems of working in compressed air arise principally in and following decompression after work. During work, nitrogen from the air is slowly dissolved in the blood stream and in some body tissues. If it is released too quickly bubbles may form, interfering with circulation. Most frequently they appear in the joints, within an hour or two of decompression, causing pain called 'the bends', but sometimes the heart or some other organ may be affected. Prevention is largely ensured by strict regulation of working times and decompression rates and times in accordance with accepted tables, but cases of compressed air illness do occur and can almost always be successfully treated by recompression under medical supervision. A third long-term effect, diagnosed by X-ray examination, is known as bone necrosis, which may be symptomless, or may eventually affect joints. It is believed that current decompression tables are effective in reducing the incidence of this to very small numbers, and that early detection by routine X-ray examination can further reduce serious cases.

Work in compressed air is subject to statutory rules, issued in 1958,[15] but modern practice is to give further protection by following the more restrictive *Medical code of practice for work in compressed air*.[16] The recommended decompression tables are known as the 'Blackpool tables'.

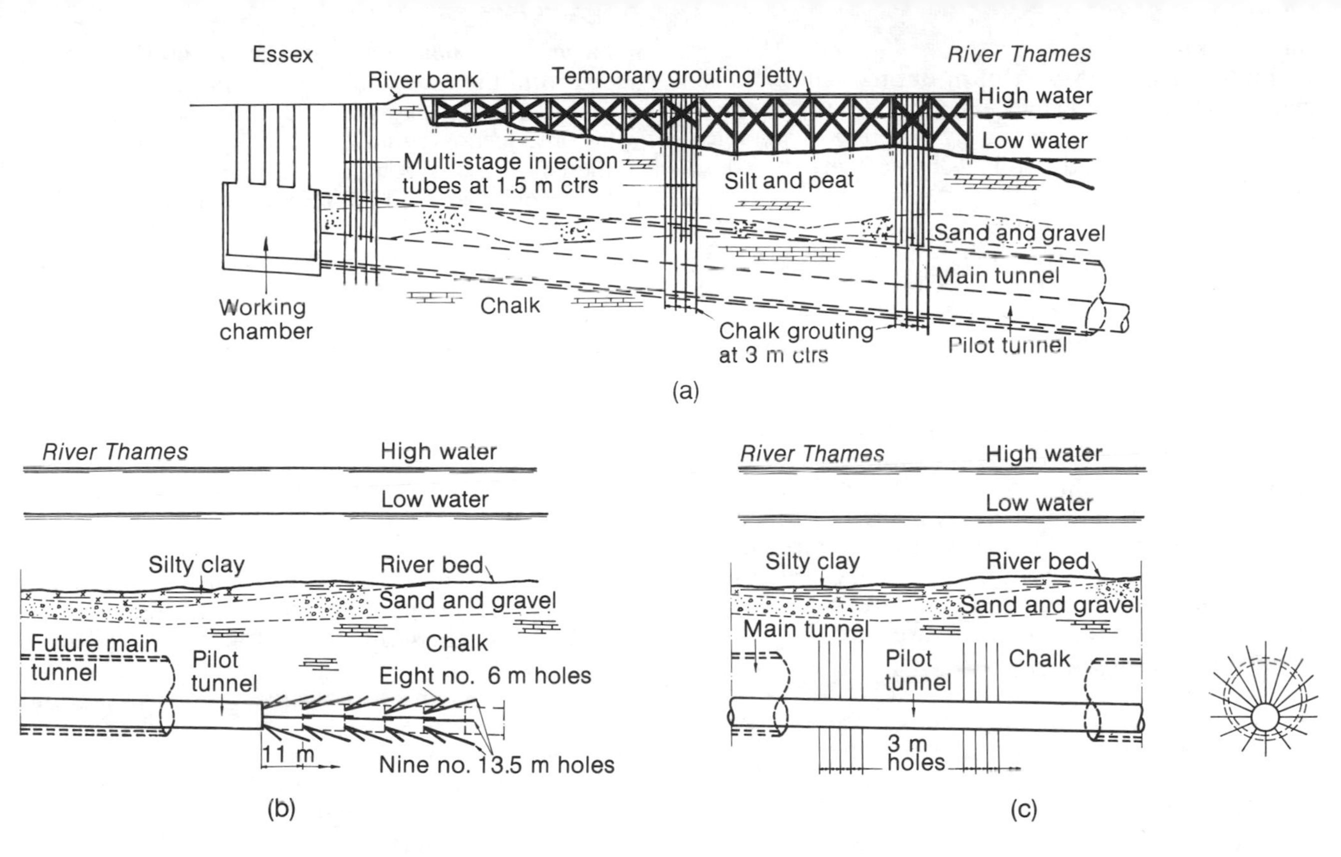

Fig. 14. Three arrangements for ground treatment below river, for second Dartford Tunnel: (a) under foreshore, from surface by use of jetty; (b) in a conical array, from face of pilot tunnel, which was constructed in 11 m lengths; (c) radially, from pilot tunnel preparatory to excavation of main tunnel

Ground treatment

Ground treatment, in advance of or during tunnelling, is directed towards strengthening the ground and reducing its permeability. This is achieved by injection of suitable grout into the pores of the soil, or by freezing the pore water.

The process of grout injection is most suitable for granular soils whose permeability allows the selected grouting fluid, pumped at a prescribed pressure into drilled holes, to penetrate the surrounding ground adequately. Cement grout is not a true chemical solution but comprises solid cement particles, of diameter about 30 μm average but up to about 100 μm, suspended in water. It will penetrate only rather coarse-grained gravels and is not suitable for fine sands or silts. A mixture with bentonite clay improves penetration and reduces permeability. The bentonite has thixotropic properties, whereby it becomes fluid when stirred but sets to a gel when undisturbed.

For finer-grained sands, chemical grouts of low viscosity can be used either as a single fluid mixture with selected viscosity and setting time, or as a two-shot process in which mixing, followed by setting, takes place in the ground. For the very finest sands, expensive resin grouts of very low viscosity may be used, but spacing of holes must be close and injection rates are necessarily slow. It is almost impossible to treat beds of silt effectively because of their fine grain structure, particularly if they form part of a mixed succession of strata, where any grout injected will tend to escape through more open soils.

Preparatory grouting for a tunnel may be done in vertical or inclined holes from the ground surface, or may be done radially from a pilot tunnel. Another method is by injection ahead from the face of a tunnel in progress, but this requires alternation of grouting and tunnelling operations, with consequent very slow advance. Fig. 14 shows how all of these methods were utilised successively in construction of the second Dartford Tunnel. Although control of water by reduced permeability is most often the prime objective, such grouting may also be adopted to reduce air losses in compressed air working. Increase of the cohesive strength of the soil is a further objective, which may be secondary to impermeability or may be primary. In fissured ground the sealing of fissures against water may be effected by suitable techniques. Except for very limited applications all these grouting techniques should be under the direction of specialists.

Freezing

Ground freezing can serve similar purposes in stopping inflow of water and providing cohesive strength in the ground. It is sometimes considered to be unduly expensive but this is not necessarily the case. Freezing is effected by driving or boring an array of tubes within which refrigerated brine or other agent is circulated. A cylinder of ground of about 1 m diameter is frozen solid over a period of a few weeks, and thereby a curtain of frozen ground is established, impermeable to water and having substantial cohesive strength. One major advantage of freezing is in treatment of silt beds, which are rendered impermeable without separate treatment. Freezing may be effected by ordinary refrigeration plant which cools down a brine circulation system; an alternative method is to use liquid nitrogen, at its extremely low temperature, where suitable supplies in tankers are available. If gravel to be excavated is frozen, concrete-breaking tools may be required or even use of explosives. Problems to be considered in use of freezing are any hazard to adjacent water mains and possible risks of frost heave. The presence of flowing water in the ground may prolong the time required for effective freezing. Also if the groundwater is saline, whether naturally or from brine leakage, a lower temperature and longer time are required.

Pipe arrays for freezing may be of any of the patterns described for injection grouting: vertical or inclined from the surface, radially from a pilot tunnel, or ahead of a tunnel face.

Groundwater lowering

Control of groundwater is a prime objective both in compressed air working and in ground treatment. Lowering of the water table in granular soils may be effected by means of well points, deep well pumping, or underdrainage below the tunnel invert.

Well points comprise a ring of tubes sunk to a depth not exceeding about 6 m, by driving or by wash-boring. They are spaced at about 1 m and connected through a ring main to suction pumps. The bottom of each tube is perforated and fitted with a strainer to exclude sand. Continuous pumping, over a period of days or weeks, draws down the water surface in a 'cone of depression' round each well point and these cones merge so that water level within the whole enclosed area is lowered. The limitation of depth arises from the maximum effective suction head

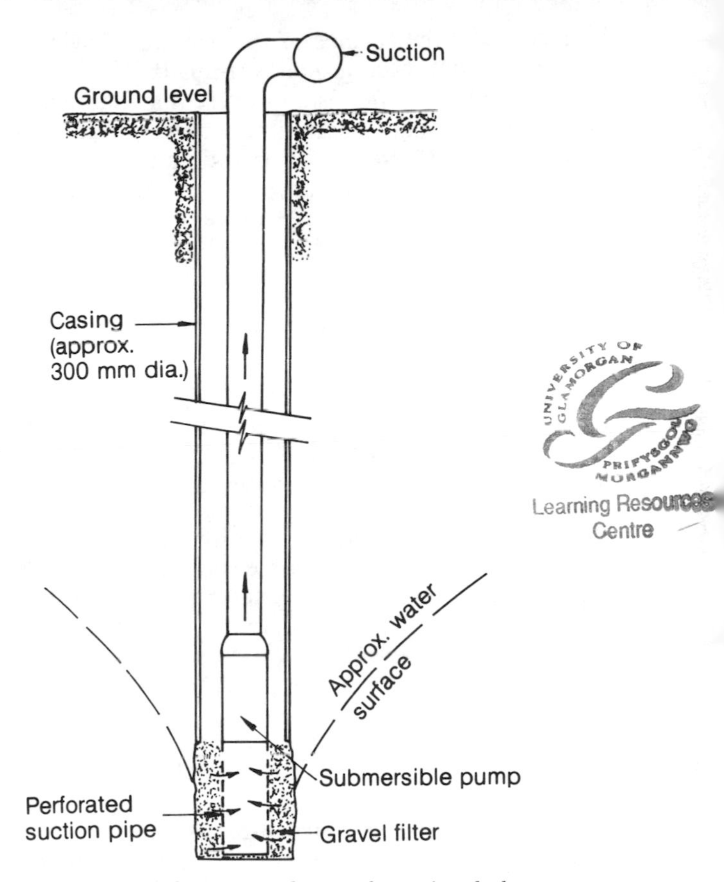

Fig. 15. Deep well for groundwater lowering below about 6 m

in a pump at the surface of the ground. Obviously this system is only useful in shallow tunnels.

Deep well pumping systems overcome the limitation in depth by utilising submersible pumps lowered to the bottom of lined bores of 300 mm diameter or more (Fig. 15). These are fitted with properly graded filters to admit water and exclude sand. The principle of drawing down the water table is exactly the same, but the individual wells are much more costly both in sinking and in pumping equipment. Spacing is likely to be 3 m or more, and the bottom of the well must be correspondingly further below the required working level. Even this system is only applicable to relatively shallow tunnels, and over a limited length, because continuous extension of the array of wells as the tunnel progresses is not likely to be acceptable either in time or in cost.

Underdrainage by means of a deep sump at a working shaft, possibly combined with a low level pilot drainage tunnel, is a system used in some subaqueous tunnels. The method is of course as appropriate to rock tunnelling as to soft ground. The Japanese Seikan Tunnel is a notable example.

10. ROCK TUNNELLING

Types of rock and rock structure

Varieties of rock are so numerous that only a broad description of leading types is possible here. The borderline between soft ground and rock for present purposes is where pick and shovel, or their modern equivalent, must give way to use of explosives.

Geological terminology is essential to description of rocks and their structure as encountered in a tunnel. The broad classification based on the mode of formation is into sedimentary, igneous and metamorphic rocks.

Sedimentary rocks, originally laid down horizontally under water and subsequently consolidated and cemented together, include sandstones, shales, mudstones and limestones. Beds may be thin and variable, or massive and relatively homogeneous. They are normally traversed by systems of vertical joints, whose spacing and characteristics govern the way in which the rock breaks up and the degree to which it is self-supporting when excavated. Shales in particular tend to fragment readily and to require much more support than does a massive sandstone. Limestones are typically well jointed: they may be thin, alternating with shales and sandstones, or thick and massive. Their solubility in acidic water can result in enlargement of fissures into vast caverns through which water flows.

Sandstones are in origin shallow-water sediments. They are frequently massive and very uniform, but may be 'current-bedded', the arrangement, sequence and grading of beds being irregular. Sandstones contain a high proportion of quartz grains, which are hard and abrasive.

Igneous rocks of molten origin range from 'plutonic' rocks, such as granite, intruded at great depth and having a coarse crystalline structure developed during slow cooling, to surface lavas, of which basalt is typical with its fine crystalline texture. Pyroclastic material is that which has been ejected from a volcano through the atmosphere and deposited on the surface; it includes volcanic ash or 'tuff'.

In massive granites the main tunnelling problem is that of breaking out the rock, which is largely or entirely self-supporting. Basalts are less uniform, and usually in thinner beds, requiring more varied techniques of excavation and support. In all cases deep penetration of weathering over long periods may cause disintegration of the crystalline structure. Pyroclastic deposits are likely to be variable and to lack consistent cohesive properties, demanding therefore continuous study of requirements for support.

Metamorphic rocks are usually ancient rocks of sedimentary or igneous origin, which have been substantially modified by heat, pressure and deformation. Tunnelling types will range from massive gneiss, resembling coarse granite, to mica schists, cleaving readily on parallel planes. In general, metamorphic rocks are of varied texture and embody discontinuities caused by stress and deformation. They are typical of mountain folding, and, where relatively new, may not be fully consolidated, so that bands of shattered rock or squeezing rock may be encountered, and physical or chemical instability may develop because of the changes caused by excavation.

Bedding and jointing

Bedding and jointing are important as constituting planes of weakness encountered in tunnelling. Bed joints are those, originally horizontal, separating changes in lithology. They are defined in terms of 'dip', the angle of slope from horizontal, and 'strike', the orientation of a level line lying in the plane of the bed joint. Jointing, within a bed, or stratum, is normally at right angles to the plane of the bed, and may be characterised by a principal set of joints, in parallel planes and more or less uniformly spaced, intersected by one or more minor sets of joints.

In driving a tunnel, the direction of drive relative to the dip and strike and the joint pattern will determine the way in which the rock in roof, sides and face tends to fracture (Fig. 16). Much depends on the cohesion in bed and other joints and on the presence of water, sometimes functioning as a lubricant.

Also of importance is the presence of a 'fault', the geological term for relative shearing displacement of the strata. The most usual form is that where the strata, on one side of a near-vertical plane, have dropped down relative to the strata on the other side. The 'throw' may be a few centimetres, or may be metres. Adjacent rock is inevitably distorted, and a band of rock may be shattered; a gap may be formed, and infilled with soft material, known as gouge.

The subject of rock mechanics deals with the behaviour under stress of rock and rock structures. To provide the necessary data for prediction the first requirement is a series of core borings, from which the detailed characteristics of the rock can be ascertained, in conjunction with as accurate a survey as practicable of geological structure and joint patterns. The possible action of water in lubricating joints and in washing out fine material must not be neglected.

Drill-and-blast cycle

Advance of a rock tunnel is a cyclic operation following

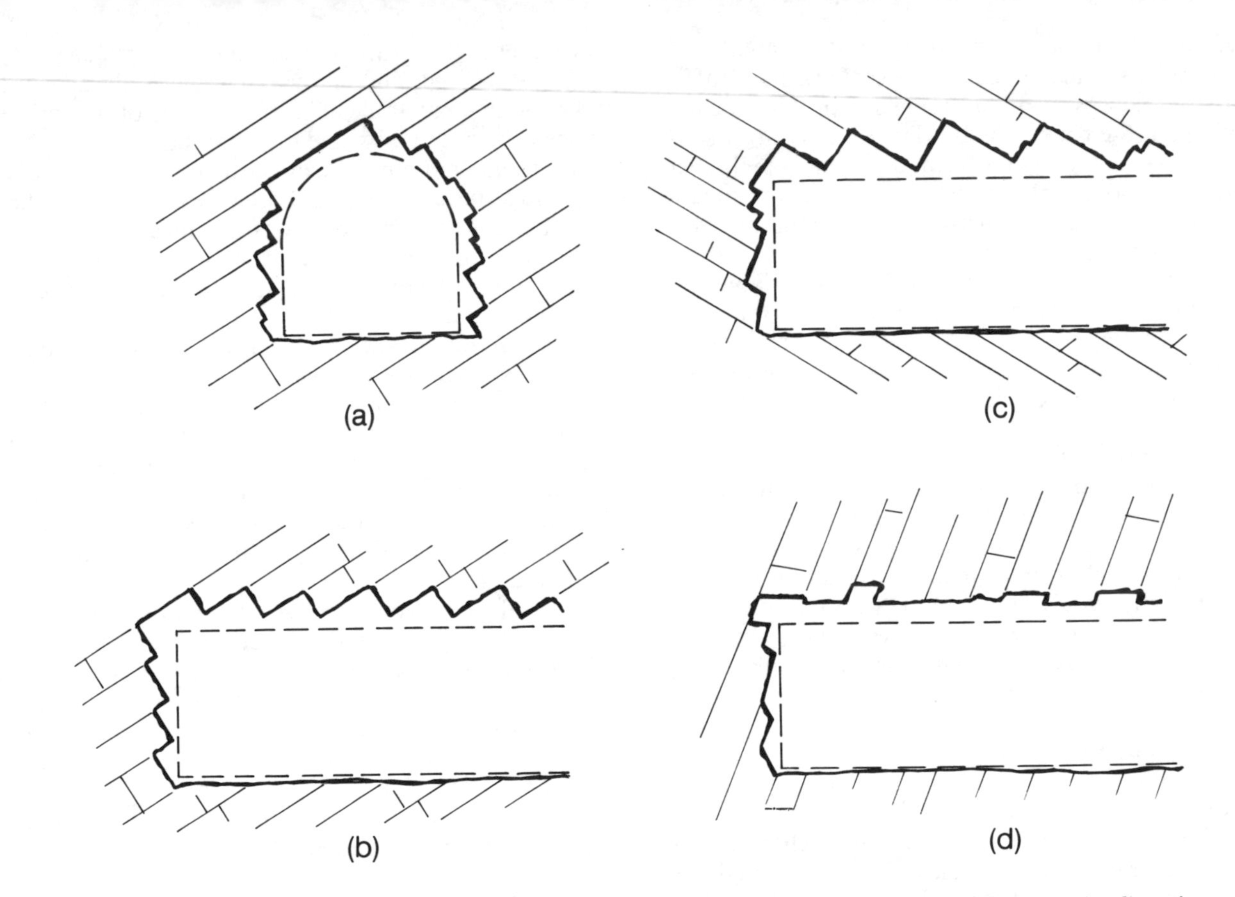

Fig. 16. Dip of strata in relation to tunnel excavation: (a) dip across tunnel; (b) tunnel following in direction of dip; (c) tunnel driven against direction of dip; (d) tunnel driven with near-vertical dip

Fig. 17. Steel arch ribs

the sequence drill face, charge, detonate, ventilate, bar down, muck out, support. There is an optimum length of advance, or 'pull', for any tunnel dependent on the nature of the rock, the safe length and timing for supports, the system of muck disposal, and the length of a shift. The tunnel roof is supported naturally by arching action across its width and forwards. Blasting destroys the support at the face and initiates a change in the stress pattern above, which may reach stability quickly or may so develop as to result in continuing rock falls. A proper appreciation of these factors determines the choice of support system, most usually in the form of steel arch ribs, packed with timber, or otherwise, to provide uniform bearing (Fig. 17). Until the spoil from blasting has been cleared supports cannot be fixed, and therefore the capacity of the disposal system must be adequate to handle

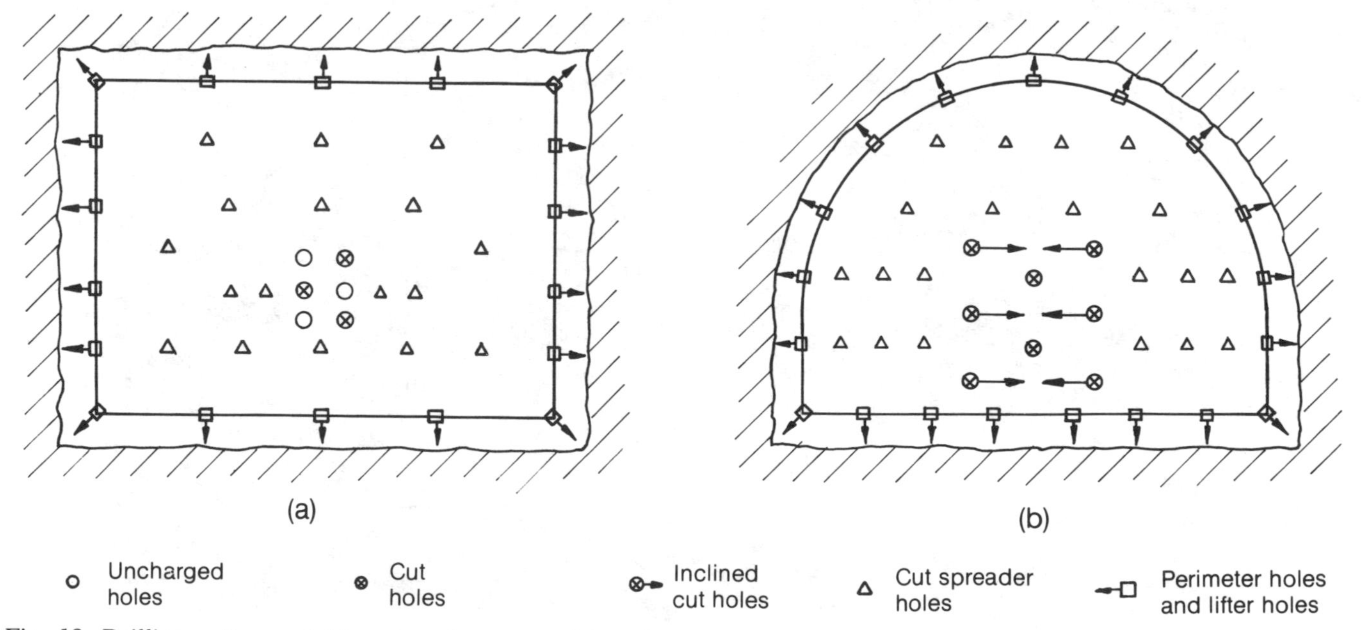

Fig. 18. Drilling patterns: (a) box type burn cut; (b) wedge cut

the volume of material, proportional to length of pull. The reference to length of shift concerns the advantage of ensuring that the changeover occurs consistently at a convenient point in the cycle. In smaller tunnels excavation is likely to be on the full face, but in larger tunnels it is often found advantageous to advance first the upper half of the face, leaving a bench, and to follow by excavating the bench as a separate phase of the cycle.

Drilling

The drilling of a pattern of holes is by means of pneumatic or hydraulic rock drills, which in a small tunnel may be lightweight tools, hand-held but with a supporting leg, or for a large face may be a battery of heavy 'drifters' mounted on a 'jumbo' drilling gantry. The depth of drilling is likely to be 2 m or more. Holes are drilled horizontally forward in a pattern designed to suit the rock and to provide convenient fragmentation. They comprise (Fig. 18) cut holes, to blast an initial cavity; cut spreader holes; perimeter holes, to trim the tunnel to the required profile; lifter, or floor, holes, immediately above floor level; and possibly easer holes, which are left uncharged, and are drilled to help in controlling the breakage.

Charging and firing

The holes are charged with appropriate quantities of explosive and each is fitted with an electrically ignited detonator. The detonators and firing circuits are so

Fig. 19 (right). Travelling shutter for in situ reinforced concrete lining

arranged that charges are fired successively and not simultaneously. All men are withdrawn to a safe distance before firing, which is the responsibility of a named man. After firing, toxic fumes must be extracted, or amply diluted, by the ventilation system before men may return to the face, where the first operation is inspection and barring down of any dangerously loose rock. This is followed by mucking out, and installation of supports. Safe storage, handling and use of explosives are described in BS 5607.[14]

Support systems

The use of steel arches is described above for any immediate support required. Sprayed concrete, additionally or in substitution, is a useful system, described in section 11. Another support technique is the installation

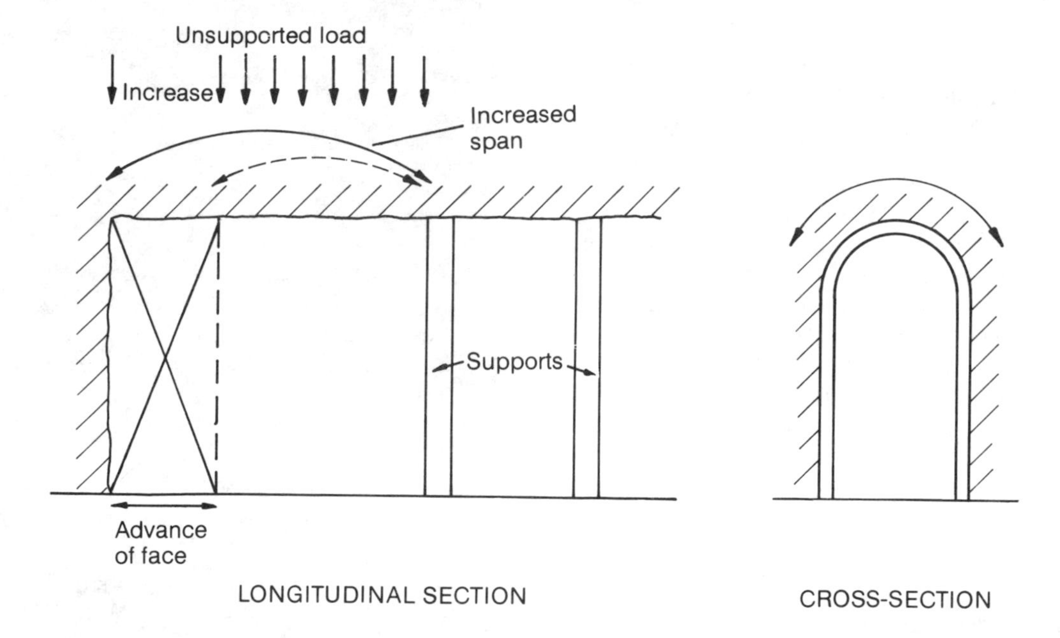

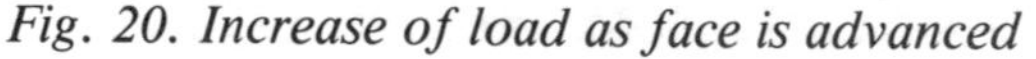
Fig. 20. Increase of load as face is advanced

of rock bolts (section 11), either to a systematic pattern in the roof, or wherever the stability of the rock is doubtful.

Permanent support is usually by in situ concrete lining. It may follow far behind excavation, as a separately organised operation. It requires lengths of carefully planned collapsible shuttering (Fig. 19) and often the use of concrete pumps or concrete placers.

The term 'stand-up time', described for soft ground tunnelling, is used also for rock. It is sometimes called the 'bridge action period'. As for soft ground, it is a function of rock type and structure in combination with unsupported span and support technique (Fig. 20). Times ranging from 20 minutes on a span of 0.8 m in very friable rock to 20 years on a span of 4 m in solid rock have been tabulated.

The assessment of rock quality and its correlation with behaviour in a tunnel has been the subject of much study, but it remains very difficult to secure any objective system of general application.

Tunnel boring machines

As in the case of soft ground tunnelling, machines fall into two main groups: full face rotary machines and independent cutting machines.

The full face rotary cutters are somewhat similar to but more powerful than soft ground machines but do not require a full shield. They are usually driven by hydraulic or electric motors acting through annular gearing. In hard rock, conical disc cutters or roller cutters are usually more effective than drag bits. The machine must be anchored within the bore so as to take the torque reaction from the cutting action, and also so as to provide forward thrust against the face. Some form of overhead canopy behind the cutter head is usually required as protection against rock falls. Such a machine is illustrated in Fig. 21. There may also be provision for drilling and fixing roof bolts close behind the face, or for spraying concrete.

Rotary cutting in some rock forms a smooth circular bore which lends itself to precast segmental lining as in soft ground.

The independent cutting machines may be boom cutters (Fig. 22) or bucket excavators. Their reactive forces are likely to be provided by the basal mounting. The shape of hole formed will be less regular, and support similar to that for drill-and-blast tunnelling will be required.

The continuous replacement of cutters necessary in hard or abrasive rock is a significant item in the cost of excavation, but there have been steady improvements in technique and materials. As with soft ground tunnelling, the prerequisite for economic use of a tunnel boring machine is that there should be a long drive in reasonably uniform conditions, and that discontinuities and emergencies outside the capacity of the machine should be avoided.

Cavern storage

The construction of large underground caverns for storage or for housing hydroelectric generating equipment extends tunnelling methods into a field beyond that of ordinary tunnels of passage. Security, economy and amenity are the prime objectives of such caverns. Apart from power stations, their widest use is probably for storage of oil, but in favourable ground conditions they

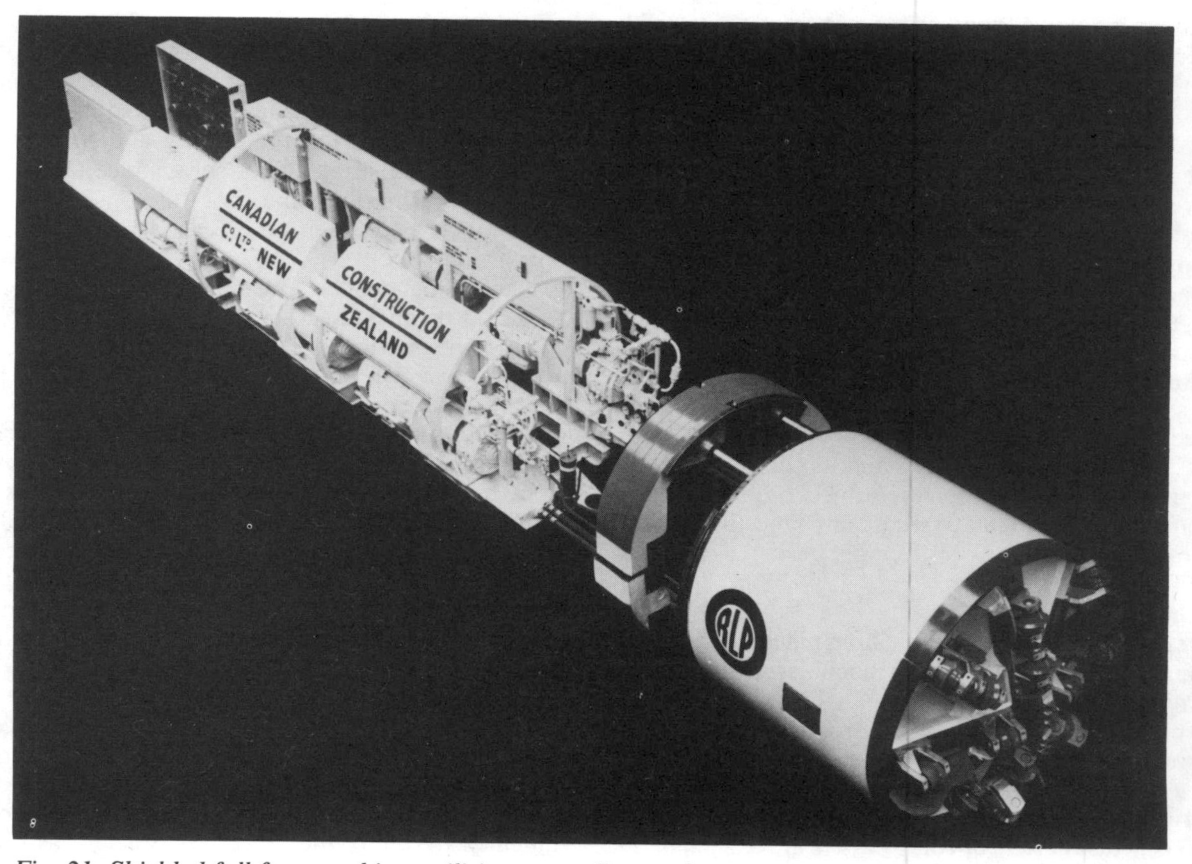

Fig. 21. Shielded full face machine, utilising expanding anchor ring for thrust reaction in hard ground or lining segments in soft; cutting head can accept disc or pick cutters (Robert Priestley Ltd)

Fig. 22. Track-mounted boom tunnelling machine with spiral cutting head; debris is swept back automatically through the machine on to the belt conveyor behind (Thyssen (GB) Ltd)

are used for such diverse purposes as cold storage, sewage treatment, gas storage, pumped energy storage (water or compressed air), military storage and disposal of nuclear waste.

As compared with the tunnelling already described, the special features and requirements are the large size of the caverns and the minimal use of lining and added support. Because of these factors, it becomes of importance to ensure in advance detailed study of the geology of the whole space involved, and investigation of rock properties and structures in terms of rock mechanics. Massive rock, free from faults or other discontinuities, is to be sought.

Excavation is likely to be with explosives, with particular care being taken to cut precisely to the desired profile with minimal disturbance of the rock structure outside the perimeter. The main load must be carried by arching of the remaining rock, possibly assisted by rock bolting and sprayed concrete. Monitoring of behaviour during construction may be quite elaborate and can give valuable guidance. In oil storage in unlined rock it is important that an effective water table is maintained to above roof level so that the lighter oil is contained by surrounding water.

11. ANCILLARY OPERATIONS IN ROCK

Rock bolting

Rock bolts are used principally in the roof of a tunnel to give added strength. In most cases the bolts are stressed in tension, but occasionally may function as unstressed dowel rods; the use of timber is sometimes preferred for temporary support if the rock is to be cut out by machine. Bolt holes are drilled up into the rock, normally at right angles to the bedding planes. The bolts (Fig. 23) are anchored at the end by a wedge or other expanding device, and are tensioned against the face of the rock by a plate and nut. Unless temporary, the bolt is grouted in, usually with cement grout, but in some circumstances with epoxy resin.

The first, obvious, action of rock bolts is to hold up insecure blocks of rock, but an even more valuable function is that the compressing of layers across the joints integrates the rock elements and allows beam action and arching action to develop.

Bolts are generally 25 mm or more in diameter and 2 m upwards in length. A pattern of bolts, at a spacing of 1 m or more, is designed to suit the ground. Adequate length to ensure that the end anchorage is in sound rock is, of course, essential.

Sprayed concrete

The use of sprayed concrete, applied directly to the newly exposed and trimmed surface of the rock in a tunnel, is a method of providing immediate and continuing support. It is an essential feature of the New Austrian Tunnelling Method (NATM). Concrete, mixed from aggregate up to about 25 mm and cement, is sprayed under pressure from a 'gun'. Water in the mix may be added initially or at the point of ejection. The mix must be carefully and accurately proportioned with close control of the water/cement ratio.

The cement grout is forced into fissures and fine cracks in the rock face, and a layer of concrete is built up over the face to any desired thickness. The effect of this is to minimise movement and disintegration of the rock, somewhat shattered by blasting, and to integrate rock and concrete so that arching action may develop and provide support. The flexibility of the thin layer of concrete allows it to accommodate to the strains and deflections in the rock without developing large bending moments causing cracking. Reinforcing steel may be incorporated in the concrete. The skill and judgement of the operators are important in effective application and in minimising losses of material as 'rebound'.

12. CUT-AND-COVER TUNNELS

The essence of the cut-and-cover method is excavation in trench from the surface, building of the tunnel structure, and backfilling over, with restoration of the surface. The method is appropriate to shallow structures where the surface can be occupied as a working site (Fig. 24). The most extensive use of cut-and-cover is for metros, of which London's Metropolitan and Metropolitan District Railways were the prototypes. Where street obstruction, inevitable during building, can be tolerated the method offers substantial advantages of cost and speed of construction, provided that depth is shallow. Another advantage of shallow depth is in operation, when access to the system from street level is quicker and easier.

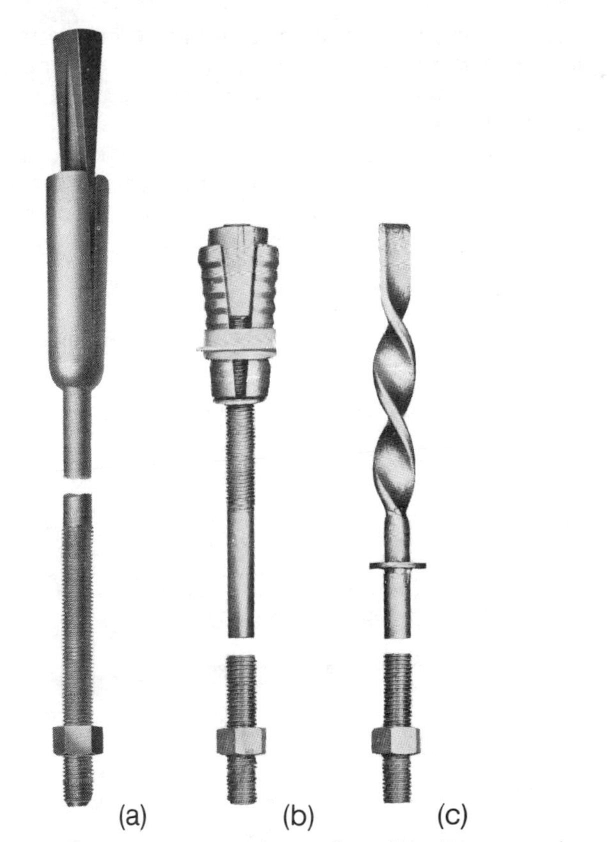

Fig. 23 Rock bolts: (a) slot and wedge; (b) expansion shell; (c) chemical grout ((b) and (c) Torque Tension Ltd)

Cut-and-cover methods are also employed in transition lengths between open-cut approaches and bored tunnels, particularly for subaqueous river crossings, and also at the portals of mountain tunnels.

Construction methods are essentially those for deep excavation in any form of construction. Trenches formed with vertical cuts in soft ground may be supported on both sides by sheet piling with walings and struts, or ground anchors. Other techniques include use of 'king piles', concrete walls cast in slurry-filled trenches, and contiguous or intersecting ('secant') bored piles. The whole width of tunnel may be built in a single trench, or side walls may first be built in separate narrow trenches (Fig. 25), followed by a wide but shallow trench for the roof slab; this allows the roadway above to be restored, and below the roof slab the dumpling may be excavated subsequently. Variants of

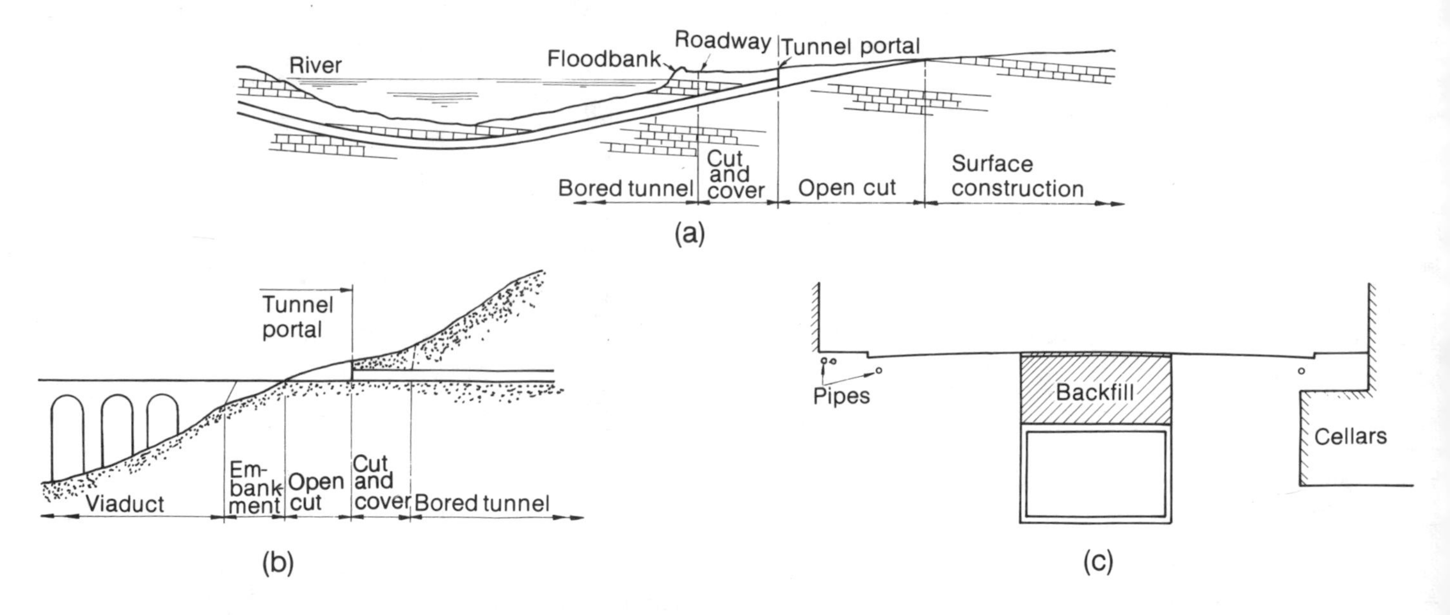

Fig. 24. Principal uses of cut-and-cover: (a) subaqueous; (b) mountain; (c) urban

these methods are possible, such as the 'umbrellas' to carry the street traffic used in the reconstruction of Oxford Circus and Bond Street stations of London's Underground.

13. SUBMERGED TUBES

For water crossings the use of prefabricated tunnel units laid and jointed in an underwater trench is an alternative to bored construction. Units upwards of 90 m in length are built elsewhere and floated to site where they are sunk into position. Two types have been developed. In the USA, a circular section is generally adopted, usually based on a steel shell fabricated by shipyard methods, concrete and reinforcement being added to complete the structure and provide the necessary weight to sink the unit. This and other fitting out may be done at a jetty. Joints between units are made initially under water as lap joints in projecting steel flanges. The tunnel rests on a carefully prepared sand or gravel bed in the trench, the sides being backfilled and the top protected against shipping and other hazards by a cover of rock fill. In the Netherlands a rectangular concrete section is normal, following the pattern adopted for the Maas Tunnel in Rotterdam (1941). The units are usually cast in a dry dock, which may be specially made for the purpose. Watertightness is ensured by use of a membrane, which may be welded steel plate or may be

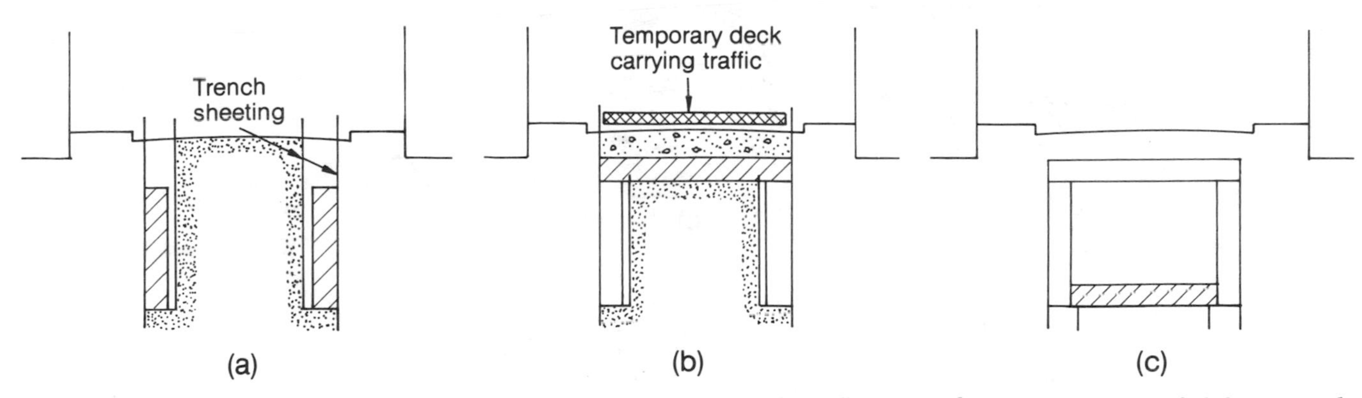

Fig. 25. Cut-and-cover in street—normal sequence: (a) stage 1—side walls in trench; (b) stage 2—roof slab in trench followed by backfill over and road reinstatement; (c) stage 3—dumpling excavated and floor concreted

bituminous. Techniques of making the concrete watertight without a membrane have been devised. When ready, the dock is flooded and the units are floated out. Jointing of the units is usually effected by use of a rubber seal, subsequently backed by welded steel plates and concrete. Units are designed to have a small positive sinking weight, and are lowered into the trench from pontoons or other floating craft on to prepared sills. A sand foundation over the whole area is then provided by pumping in suitably graded sand in water. The sides and top are backfilled as before.

It will be obvious that the techniques are totally different from those of bored tunnelling. Shipyard and concrete fabrication methods are employed, and handling and placing demand skill in using tugs and anchors and in coping with river or tidal currents.

The advantages in suitable conditions are that a shallower depth to road level is possible (Fig. 26) because

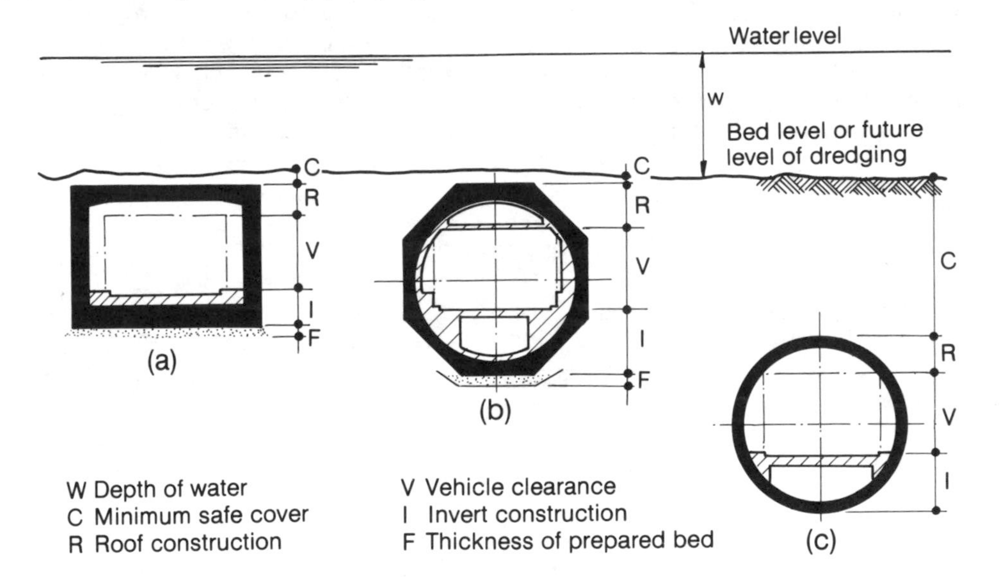

Fig. 26. Comparative depths required for tunnels of similar function but differing forms of construction: (a) submerged concrete tunnel; (b) submerged steel tunnel; (c) bored tunnel (shield)

much less protective cover is necessary than in boring a tunnel and, with the rectangular form, the construction depth is also reduced. Resources of labour and skill may also be more readily available in the area concerned.

14. COSTS

The cost of tunnel construction must first be considered in relation to the value of the service it is intended to provide, and for this purpose reasonably accurate estimates must be prepared by the Engineer. These estimates will be refined as the design of the project is determined, up to the stage of inviting tenders from contractors. Those tendering are obliged to commit themselves to fixed prices, except insofar as provisional and contingency items are allowed either explicitly or in interpretation of the terms of the contract. For example, there might be provision for the cost of compressed air working if found essential, and there is frequently extra payment to be assessed in respect of ground conditions which could not reasonably have been foreseen.

The preparation of estimates of this kind, up to the letting of a contract, is a matter requiring experience, judgement and detailed assessment of the costs likely to be incurred.

Actual costs of previous tunnels provide initially a valuable basis, but the impact of continuing inflation of prices makes very uncertain any process of applying indexes to earlier figures. As work on a large tunnel may be spread over several years, and interest will be due on capital sums expended, the dates at which costs are incurred are significant both for promoter and contractor.

An estimate will thus be built up from engineering costs with inflation and financing costs superimposed as thought appropriate. The elements of an engineering estimate are the costs and period of use of plant and equipment, materials, energy consumption, labour and overheads.

A purpose-built tunnel boring machine may be the most expensive single item; it must be ordered many months ahead of delivery, and may have to be written off against the single job unless there is a reasonable prospect of its further use. Other specialised items of equipment must similarly be obtained and costed as appropriate, possibly on a hire basis over their period of use. Materials such as steel, cast iron, precast concrete, cement, concrete aggregates and timber can be assessed on current quotations with appropriate provision for inflationary increases.

Energy consumption, whether as electricity or liquid fuel, can be calculated in terms of power and periods of use of plant (tunnel boring machines, conveyors, locomotives, hoists, compressors, hydraulic power packs etc.) and also lighting.

Labour requirements can be estimated by 'manning the job' on paper and applying the appropriate time rates, bearing in mind the problems of availability of men of the requisite skill, possibly in an area remote from industry. As an example, the manning of a tunnel drive is built up by visualising the number of miners and labourers, the attendant locomotive driver and others, and the surface workers including crane driver and banksman. Fitters, electricians and supervisory staff are then allocated. A reasonable output per shift is assumed and the estimated labour cost per unit of tunnel thus derived.

Site overheads can similarly be estimated and allocated to the various types of work.

It will be obvious that accurate estimates of the rate of advance are critical, and that any unforeseen delays can be very costly. Adequate information from boreholes and tests is essential to forming a correct judgement.

All these component costs, and others such as those for site preparation and temporary works, will be combined into an overall estimate. In a contract they are usually allocated in detail to the separate items in the bills of quantities, payment being made against actual measured quantities.

15. ACKNOWLEDGEMENTS AND FURTHER READING

Acknowledgements for illustrations

Nearly all the illustrations were originally prepared for and published in *Tunnels: planning, design, construction,* by Megaw and Bartlett, and are reproduced here with the kind consent of the publishers, Ellis Horwood Ltd, Chichester.

Permission for use of copyright photographs has kindly been given as follows:

Fig. 2 Engineering Laboratory Equipment Ltd
Fig. 21 Robert L. Priestley Ltd
Fig. 22 Thyssen (GB) Ltd.

Further reading

The classic books are mostly out of print, but, apart from their historical interest, they still provide much useful and practical guidance on tunnel construction.

1. SIMMS F.W. *Practical tunnelling*. Crosby Lockwood, London, 1844 (revised and enlarged 1859, 1877, 1896).
2. DRINKER H.S. *Tunneling, explosive compounds and rock drills*. J. Wiley, New York, 1878.
3. COPPERTHWAITE W.C. *Tunnel shields and the use of compressed air in subaqueous work*. Constable, London, 1906 (reprinted A.A. Mathews, Arcadia, California, 1968).
4. HEWETT B.M. and JOHANNESSON S. *Shield and compressed air tunneling*. McGraw Hill, New York, 1922 (reprinted A.A. Mathews, Arcadia, California, 1968).

Comprehensive books of more recent date are not numerous. Those of particular value include the following:

5. RICHARDSON H.W. and MAYO R.S. *Practical tunnel driving*. McGraw Hill, New York, 1941.
6. PEQUIGNOT C.A. (ed.). *Tunnels and tunnelling*. Hutchinson, London, 1963 (out of print).
7. SZECHY K. *The art of tunnelling*. Akademiai Klado, Budapest, 1973, 2nd edn.

This last named is very comprehensive in setting out many theoretical analyses of tunnel design, and also describing and illustrating the construction of many important tunnels.

8. MEGAW T.M. and BARTLETT J.V. *Tunnels: planning, design, construction*. Ellis Horwood, Chichester, 1981 (vol. 1) and 1982 (vol. 2).

This covers much more extensively the subject matter of this booklet. It also incorporates a lengthy bibliography including, in addition to books, a list of some 200 papers on tunnels published by the Institution of Civil Engineers, and a similar list of American Society of Civil Engineers papers.

9. HOEK E. and BROWN E.T. *Underground excavations in rock*. Institution of Mining and Metallurgy, London, 1980.

This covers the structural aspects of its subject, including cavern excavation, comprehensively and authoritatively.

10. HARDING Sir H.J.B. *Tunnelling history and my own involvement*. Golder Associates, Toronto, 1981.

In addition to its general interest there is particular value in the examples from the author's own experience setting out in some detail the actions taken to overcome difficulties.

11. CRAIG R.N. and MUIR WOOD A.M. *A review of tunnel lining practice in the United Kingdom*. Transport and Road Research Laboratory, Crowthorne, 1978.

The many varieties of UK segmental linings are described and illustrated in useful detail, together with discussion of design methods. In situ linings are also covered.

12. BARTLETT J.V. and KING J.R.J. Soft ground tunnelling. *Proc. Instn Civ. Engrs,* Part 1, 1975, **58**, Nov., 615–628; discussion 1976, **60**, Aug., 483–487.

This paper is an introduction to the subject for the young engineer, and includes discussion of preliminary planning, contract preparation and construction.

There are various statutory regulations, British Standards and other publications, mostly applicable to works construction generally. A few are particular to tunnelling, including:

13. BRITISH STANDARDS INSTITUTION. *Code of practice for safety in tunnelling in the construction industry*. BSI, London, 1982, BS 6164.
14. BRITISH STANDARDS INSTITUTION. *Safe use of explosives in the construction industry*. BSI, London, 1978, BS 5607.
15. *Work in compressed air: special regulations*. Her Majesty's Stationery Office, London, SI 1958 no. 61.
16. CONSTRUCTION INDUSTRY RESEARCH AND

INFORMATION ASSOCIATION. *Medical code of practice for work in compressed air*. CIRIA, London, 1982, report 44 (3rd edn).

17. *Tunnels and Tunnelling*. Monthly periodical providing up-to-date information. Supplied to all members of British Tunnelling Society.